ANSI C2-1981
Revision of
C2.1-1971
C2.2-1976
C2.3-1973
C2.4-1973

American National Standard

National Electrical Safety Code

1981 Edition

Approved August 15, 1980
American National Standards Institute

Secretariat

Institute of Electrical and Electronics Engineers, Inc

Published by

Institute of Electrical and Electronics Engineers, Inc
345 East 47th St, New York, N.Y. 10017

An American National Standard implies a consensus of those substantially concerned with its scope and provisions. An American National Standard is intended as a guide to aid the manufacturer, the consumer, and the general public. The existence of an American National Standard does not in any respect preclude anyone, whether he has approved the standard or not, from manufacturing, marketing, purchasing, or using products, processes, or procedures not conforming to the standard. American National Standards are subject to periodic review and users are cautioned to obtain the latest editions.

CAUTION NOTICE: This American National Standard may be revised or withdrawn at any time. The procedures of the American National Standards Institute require that action be taken to reaffirm, revise, or withdraw this standard no later than five years from the date of publication. Purchasers of American National Standards may receive current information on all standards by calling or writing the American National Standards Institute.

Abstract

This standard covers basic provisions for safeguarding of persons from hazards arising from the installation, operation, or maintenance of 1) conductors and equipment in electric-supply stations, and 2) overhead and underground, electric-supply and communication lines. It also includes work rules for the construction, maintenance, and operation of electric-supply and communication lines and equipment.

The standard is applicable to the systems and equipment operated by utilities, or similar systems and equipment, of an industrial establishment or complex under the control of qualified persons.

This standard consists of the definitions, grounding rules, and Parts 1, 2, 3 and 4 of the 1981 edition of the National Electrical Safety Code.

Key words: Communication industry safety; construction of communication lines; construction of electric-supply lines; electric-supply stations; electric utility stations; electrical safety; high voltage safety; operation of communication systems; operation of electric-supply systems; power station equipment; power station safety; public utility safety; safety work rules; underground communication line safety; underground electric line safety.

Foreword

(This Foreword is not a part of American National Standard, National Electrical Safety Code, C2.)

This publication consists of the parts of the National Electrical Safety Code (NESC) currently in effect. The former practice of designating parts by editions has become impractical. In the 1977 edition, Parts 1 and 4 were 6th Editions, Part 2 was a 7th edition, Part 3, a revision of the 6th edition, Part 2, section 29, did not cover the same subject matter as the 5th edition, Part 3 (withdrawn in 1970). In this 1981 edition revisions have been made in all parts. It is recommended that reference to the National Electrical Safety Code be made solely by the years on the published volume and desired part number. Separate copies of the individual parts are not available.

Work on the National Electrical Safety Code started in 1913 at the National Bureau of Standards, resulting in the publication of NBS Circular 49. The last complete code (the fifth edition, NBS Handbook H30) was issued in 1948, although separate portions had been available at various times starting in 1938. Part 2, Definitions, and the Grounding Rules, sixth edition, were issued as NBS Handbook H81. ANSI C2.2-1960, in November 1961, but work on other parts was not actively in process again until 1970. In 1970 the C2 committee decided to delete from future editions, Rules for the Installation and Maintenance of Electric Utilization Equipment (Part 3 of the fifth edition) now largely covered by the National Electrical Code (ANSI C1), and the Rules for Radio Installations (Part 5 of the fifth edition). The Discussion of the National Electrical Safety Code, issued as NBS Handbook H4 (1928 edition) for the fourth edition of the NESC, and as NBS Handbook H39 for Part 2 of the Grounding Rules of the fifth edition, was not published for the sixth edition. The major change in the 1977 edition was the extensive revision of Part 2, Overhead Lines.

The 1981 Edition of the National Electrical Safety Code has been revised to update all references, incorporate the rules common to all parts into Section 1, and all definitions (now coordinated for all parts) into Section 2. Sections 1, Introduction, 2, Definitions, and 9, Grounding Methods, are applicable to each of the Parts 1, 2, 3, and 4.

Section 9, Grounding Methods, includes major clarifications of the Rules 92, Point of Connection of the Grounding Conductor and the addition in 94B5 of a specification for semiconducting concentric neutral cable jackets.

Part 1, Electric Supply Stations, has been revised throughout; major changes include new or revised rules concerning illumination, extension of clearance tables to include circuits operating at a maximum of 800 kV, working space about equipment, hazardous locations (extensive additions), rotating equipment, storage batteries, switchgear and metal enclosed bus.

Part 2, Overhead Lines, has extensive minor clarifications throughout the part. Major revisions have been made in Rule 233. Clearance Between Wires, Conductors, or Cables, carried on Different Support-

5

ing Structures; The tables of the Electrical Component of Clearance for the higher voltages have been simplified by omitting some values (they are primarily a check on calculations). For circuits with known switching surge factors, ·the lower limit is now 98 kVac/ 139 kVdc. The tables of Rule 261 have been reorganized (without change in requirement values) to be more convenient to use.

Part 3, Underground Lines has clarifications and significant changes in Section 33, Supply Cable; Rule 354C2, Random Separation, now has a clause on semiconducting jacketed grounding conductors paralleling that introduced in Section 9.

Part 4, Work Rules, has a few minor clarifications; the minimum clearance from live parts, for voltages above 230 kV, is now covered only in the table providing for consideration of the surge factor.

The Institute of Electrical and Electronics Engineers was designated as the administrative secretariat for C2 in January 1973, assuming the functions formerly performed by the National Bureau of Standards.

Comments on the rules and suggestions for their improvement are invited, especially from those who have experience in their practical application. In future editions every effort will be made to improve the rules, both in the adequacy of coverage and in the clarification of requirements. Comments should be addressed to:

Secretary
American National Standards Committee C2
Institute of Electrical and Electronics Engineers
345 East 47th Street
New York, N.Y. 10017

A representative Committee on Interpretations has been established to prepare replies to requests for interpretation of the rules contained in the code. Requests for interpretation should state the rule in question as well as the conditions under which it is being applied. Interpretations are intended to clarify the intent of specific rules and are not intended to supply consulting information on the application of the code. Requests for interpretation should be addressed to:

Secretary — Interpretations
National Electrical Safety Code Committee, C2
IEEE Standards Office
345 East 47th Street
New York, N.Y. 10017

If the request is suitable for processing, it will be sent to the Interpretations Committee. After consideration by the committee, which may involve many exchanges of correspondence, the inquirer will be notified of its decision. Decisions will be published from time to time in cumulative form and may be ordered.

The code as written is a voluntary standard. However, some editions and some parts of the code have been adopted, with and without changes, by some state and local jurisdictional authorities. To determine the legal status of the National Electrical Safety Code in any particular state or locality within a state, the authority having jurisdiction should be contacted.

Standards Committee Membership

At the time this standard was approved, American National Standards Committee C2 had the following membership:

R. H. Lee, *Chairman*   **E. W. Glancy,** *Vice Chairman*
C. R. Muller, *Secretary*

Organization represented	*Name*
American Insurance Association	E. S. Charkey
	Richard Tufts (Alt)
American Public Power Association	Herbert I. Blinder
American Society of Civil Engineers	G. M. Wilhoite
	Sidney Alpert (Alt)
American Transit Association	John G. Mombach
	Herbert J. Scheuer (Alt)
Association of American Railroads	Leo M. Himmel
Association of Edison Illuminating Companies	Robert B. Bartlett
Bonneville Power Administration US Department of Energy	E. J. Yasuda
Bureau of Reclamation US Department of the Interior	(vacant)
Edison Electric Institute	D. E. Ruff
	J. Harry Beckmann (Alt)
	Matthew C. Mingoia (Alt)
Electronic Industries Association	C. D. Hansell
Institute of Electrical and Electronics Engineers	A. A. Chase
	Frank A. Denbrock (Alt)
	T. W. Haymes, Jr. (Alt)
	Vernon R. Lawson
International Association of Electrical Inspectors	L. E. La Fehr (Actg)
International Association of Governmental Labor Officials	Bernard L. O'Neil
	Edward C. Lawry (Alt)
International Brotherhood of Electrical Workers	R. W. MacDonald
International Municipal Signal Association	Edgar P. Grim
National Association of Regulatory Utility Commissioners	Allen L. Clapp
	Clarence F. Reiderer (Alt)
	Robert J. Buckley (Alt)

Standards Committee Membership

National Cable Television Association,
Inc. .... Robert D. Bilodeau
National Electrical Contractors
Association .... Gordon M. Collins
Charles J. Hart (Alt)
National Electrical Manufacturers
Association .... D. C. Fleckenstein
D. R. Smith (Alt)
National Safety Council .... Tony E. Branan
Phil Schmidt (Alt)
Public Service Commission of
Wisconsin .... C. F. Riederer
Rural Electrification Administration,
US Department of Agriculture ...... James M. McCutchen
Edward J. Cohen (Alt)
The Telephone Group .... E. W. Glancy
C. D. Hansell (Alt)
Western Area Power Administration
US Department of Energy.......... C. A. Cabral
Canadian Standards Association
(Liaison Representative) .... P. Ralston

The eight subcommittees which prepared this revision had the following membership:

SC1 Purpose, Scope, Application, and Definitions — Sections 1, 2

E. W. Glancy, *Chairman*

Charles J. Blattner	(SC3)	W. A. Thue	(Alt/SC7)
Joseph L. Mills	(Alt)	James M. McCutchen	(SC4)
F. A. Denbrock	(SC5)	Ronald H. Martineau	(Alt)
Allen L. Clapp	(Alt)	Wayne B. Roelle	(Alt)
J. P. Fitzgerald	(SC8)	Paul S. Shelton	(SC2)
E. W. Glancy	(SC6,7)	Archie W. Cain	(Alt)
Algard A. Chase	(Alt/SC6)		

8

SC2 Grounding Methods Section 9

Paul S. Shelton, *Chairman*   **Archie W. Cain,** *Secretary*

Charles D. Hansell
T. W. Haymes
Ralph H. Lee
Robert W. Macdonald

S. D. Mauney
R. R. Mortt
L. F. Murray
L. H. Sessler

SC3 Electric Supply Stations Sections 10-19

Charles J. Blattner, *Chairman*   **Joseph L. Mills,** *Secretary*

Anthony J. Bauman
Lee A. Belfore
William J. Celio
John W. Colwell
W. Johnson
Lee Julson

Russell E. Lincoln
T. A. Lucas
William J. Lyon
Robert W. Macdonald
Robert E. Penn
J. W. Richardson

Clifford W. Schmitz, Jr

SC4 Overhead Lines — Clearances Section 23

James M. McCutchen, *Chairman*

Wayne B. Roelle, *Vice Chairman*

Ronald H. Martineau, *Secretary*

Robert J. Buckley
 A. E. Zahller (Alt)
Allen L. Clapp
John A. Collins
E. W. Glancy
D. E. Hooper

Robert F. Lehmann
Robert W. Macdonald
Richard M. Mason
August C. Pfitzer
J. Lester Price
V. J. Warnock

E. J. Yasuda

SC5 Overhead Lines — Strength and Loading Sections 24-26

F. A. Denbrock, *Chairman*   **Allen L. Clapp,** *Secretary*

R. H. Beck
W. C. Currence, Jr.
 Arnold Milbright, (*Alt*)
 R. J. Mohler, (*Alt*)
W. C. Engelmann
W. F. Fuller

Donald G. Heald
H. Gordon Jensen
R. A. Kravitz
Ron Marsico
G. M. Wilhoite
 Walter Thomas, (*Alt*)

Walter A. Swanson

9

SC6 Overhead Lines General, Insulation and Miscellaneous
Sections 20, 22, 28

E. W. Glancy, *Chairman*   **Algard A. Chase,** *Secretary*

G. W. Cock
J. G. Cunningham
Kenneth R. Edwards
D. C. Fleckenstein
James M. McCutchen
D. T. Michael
John T. Shincovich

INSULATION
W. D. Archer
Alfred C. Miner
L. R. Odle
T. A. Pinkham
D. N. Rice
E. L. Shaffer

SC7 Underground Lines Sections 30-32

E. W. Glancy, *Chairman*   **W. A. Thue,** *Vice Chairman*

Billie L. Adair
William O. Andersen
E. L. Bennett
Charles C. Bleakley
Larry G. Clemons
Gordon M. Collins
  Charles J. Hart, (Alt)
Frederick S. Cook
Rafael A. Fernandez

G. M. Fitzpatrick
W. N. Fredenburg
Paul M. Henkels
F. W. Koch
Robert W. Macdonald
James M. McCutchen
John G. Mombach
W. C. Reinhardt
Lanny L. Smith

SC8 Work Rules Section 40-44

J. P. Fitzgerald, *Chairman*

James D. Beck
Tony E. Branan
C. A. Cabral
James M. Degen
A. J. Fehrenbach
Charles J. Hart
Henry J. Kientz
Robert W. Macdonald

Lewis E. Meeker
Eldon E. O'Neal
Donald W. Richmond
Russell R. Selbo
George E. Smith
Orville J. Toulouse
Joseph M. VanName
Euel Wade

E. J. Yasuda

10

SI Conversion Units

In view of present accepted practice in the United States in this technological area, common US units of measurement have been used throughout this code. In recognition of the position of the United States as a signatory to the General Conference on Weights and Measures, which gave official status to the metric SI system of units in 1960, conversion factors applicable to the US units used in this code are presented below.

Length
  1 in = 0.0254* meter (m)
  1 ft = 0.3048* meter (m)

Force (strength)
  1 lbf = 4.448 newtons

Electricity
  1 volt — SI Unit
  1 ampere — SI Unit

Power
  1 watt — SI Unit

---

*Exactly.

Letter Symbols for Units

This code uses standard symbols for units. They have the following meanings:

A   ampere
ft   foot
in   inch
kV   kilovolt (1000 volts)
V   volt

Introduction

Definitions

Grounding Methods

Supply Stations

Overhead Lines

Underground Lines

Work Rules **4**

Index **I**

Contents

Contents

Contents

Contents

18

Contents

Part 2 Overhead Lines

Contents

Contents

Contents

Contents

Contents

Contents

Contents

Contents

Contents

31

Contents

Contents

Contents

Contents

Contents

Contents

Contents

Contents

Contents

Contents

Contents

Contents

44

Contents

Section 1. Introduction to the National
Electrical Safety Code

010. Purpose

The purpose of these rules is the practical safeguarding of persons during the installation, operation, or maintenance of electric supply and communication lines and their associated equipment. They contain minimum provisions considered necessary for the safety of employees and the public. They are not intended as a design specification or an instruction manual.

011. Scope

These rules cover supply and communication lines, equipment, and associated work practices employed by an electric supply, communication, railway, or similar utility in the exercise of its function as a utility. They cover similar systems under the control of qualified persons, such as those associated with an industrial complex.

It does not cover installations in mines, ships, railway rolling equipment, aircraft, or automotive equipment, or utilization wiring except as covered in Parts 1 and 3.

012. General Rules

All electric supply and communication lines and equipment shall be designed, constructed and maintained to meet the requirements of these rules. For all particulars not specified in these rules, construction and maintenance should be done in accordance with accepted good practice for the given local conditions.

013. Application

A. New Installations and Extensions

1. These rules shall apply to all new installations and extensions, except that they may be waived or modified by the administrative authority. When so waived or modified, equivalent or greater safety shall be provided in other ways, including special working methods.
2. Types of construction and methods of installation other than those specified in the rules may be used experimentally to obtain information, if done where qualified supervision is provided.

B. Existing Installations
1. Existing installations including maintenance replacements, which comply with prior editions of the code, need not be modified to comply with these rules except as may be required for safety reasons by the administrative authority.
2. Where conductors or equipment are added, altered, or replaced on an existing structure, the structure or the facilities on the structure need not be modified or replaced if the resulting installation will be in compliance with the rules which were in effect at the time of the original installation.

014. Waiver

The person responsible for an installation may modify or waive certain rules in the case of emergency or temporary installations. When the rules are waived or modified during emergencies, the installation shall be brought into compliance with these rules after the emergency has ceased. In the case of non-emergency temporary installations, only those rules involving permanence or durability of the installation may be modified.

015. Intent

Rules in this code which are to be regarded as mandatory are characterized by the use of the word *shall*. Where a rule is of an advisory nature, to be followed insofar as practical, it is indicated by the use of the word *should*. Other practices which are considered desirable, but are not intended to be mandatory, are stated as *RECOMMENDATIONS*.

NOTES contained herein, other than footnotes to tables, are for information purposes only and are not to be considered as mandatory or as part of the code requirements.

016. Effective Date

Unless otherwise stipulated by the administrative authority, this code shall become effective 180 days following the date of its publication.

Supersedes
C2.1-1971, Section A
C2.2-1976, Section A
C2.3-1973, Section A
C2.4-1973, Section A

D

American National Standard

National Electrical Safety Code

Section 2. Definitions of Special Terms

Section 2. Definitions of Special Terms

The following definitions are for use with the National Electrical Safety Code.

For other use, and for definitions not contained herein, see ANSI/IEEE Std 100-1977*, IEEE Standard Dictionary of Electrical and Electronics Terms.

administrative authority. An organization exercising jurisdiction over application of this code.

alive or live. *See* **energized.**

automatic. Self-acting, operating by its own mechanism when actuated by some impersonal influence—as, for example, a change in current strength; not manual; without personal intervention. Remote control that requires personal intervention is not automatic, but manual.

backfill (noun). Materials such as sand, crushed stone, or soil, which are placed to fill an excavation.

ballast section (railroads). The section of material, generally trap rock, which provides support under railroad tracks.

bonding. The electrical interconnecting of conductive parts, designed to maintain a common electrical potential.

cable. A conductor with insulation, or a stranded conductor with or without insulation and other coverings (single-conductor cable) or a combination of conductors insulated from one another (multiple-conductor cable).

spacer cable. A type of electric supply line construction consisting of an assembly of one or more covered conductors, separated from each other and supported from a messenger by insulating spacers.

cable jacket. A protective covering over the insulation, core, or sheath of a cable.

cable sheath. A conductive protective covering applied to cables.

NOTE: A cable sheath may consist of multiple layers of which one or more is conductive.

cable terminal. A device which provides insulated egress for the conductors. *Syn:* **termination.**

---

*Supersedes ANSI C42.100-1972 (IEEE Std 100-1972).

circuit. A conductor or system of conductors through which an electric current is intended to flow.

circuit breaker. A switching device capable of making, carrying and breaking currents under normal circuit conditions and also making, carrying for a specified time, and breaking currents under specified abnormal conditions such as those of short circuit.

common use. Simultaneous use by two or more utilities of the same kind.

communication lines. *See* lines, communication.

conductor.
1. A material, usually in the form of a wire, cable, or bus bar, suitable for carrying an electric current.
2. bundled conductor. An assembly of two or more conductors used as a single conductor and employing spacers to maintain a predetermined configuration. The individual conductors of this assembly are called subconductors.
3. covered conductor. A conductor covered with a dielectric having no rated insulating strength or having a rated insulating strength less than the voltage of the circuit in which the conductor is used.
4. grounded conductor. A conductor which is intentionally grounded, either solidly or through a noninterrupting current-limiting device.
5. grounding conductor. A conductor which is used to connect the equipment or the wiring system with a grounding electrode or electrodes.
6. insulated conductor. A conductor covered with a dielectric (other than air) having a rated insulating strength equal to or greater than the voltage of the circuit in which it is used.
7. lateral conductor. A wire or cable extending in a general horizontal direction at an angle to the general direction of the line conductors.
8. line conductor. A conductor intended to carry electric currents, supported by poles, towers, or other structures, but not including vertical or lateral conductors.
9. open conductor. A type of electric supply or communication line construction in which the conductors are bare, covered or insulated and without grounded shielding, individually supported at the structure either directly or with insulators. *Syn:* open wire.

conductor shielding. An envelope which encloses the conductor of a cable and provides an equipotential surface in contact with the cable insulation.

conduit. A structure containing one or more ducts.

NOTE: Conduit may be designated as iron pipe conduit, tile conduit, etc. If it contains one duct only it is called *single-duct conduit;* if it contains more than one duct it is called *multiple-duct conduit,* usually with the number of ducts as a prefix, for example, *two-duct multiple conduit.*

conduit system. Any combination of duct, conduit, conduits, manholes, handholes and vaults joined to form an integrated whole.

current-carrying part. A conducting part intended to be connected in an electric circuit to a source of voltage. Non-current-carrying parts are those not intended to be so connected.

de-energized. Free from any electrical connection to a source of potential difference and from electric charge; not having a potential different from that of the earth.

NOTE: The term is used only with reference to current-carrying parts which are sometimes energized (alive). *Syn:* **dead.**

disconnecting or isolating switch. A mechanical switching device used for changing the connections in a circuit, or for isolating a circuit or equipment from a source of power.

NOTE: it is required to carry normal load current continuously, and also abnormal or short-circuit current for short intervals as specified. It is also required to open or close circuits either when negligible current is broken or made, or when no significant change in the voltage across the terminals of each of the switch poles occurs. *Syn:* **disconnector, isolator.**

duct. A single enclosed raceway for conductors or cable.

effectively grounded. Intentionally connected to earth through a ground connection or connections of sufficiently low impedance and having sufficient current-carrying capacity to prevent the build-up of voltages which may result in undue hazard to connected equipment or to persons.

electric supply equipment. Equipment which produces, modifies, regulates, controls, or safeguards a supply of electric energy. *Syn:* **supply equipment.**

electric supply lines. *See* **lines.**

electric supply station. Any building, room, or separate space within which electric supply equipment is located and the interior of which is accessible, as a rule, only to qualified persons.

NOTE: This includes generating stations and substations and generator, storage battery, and transformer rooms. *Syn:* **supply station.**

enclosed. Surrounded by case, cage, or fence designed to protect the contained equipment and minimize the possibility, under normal conditions, of dangerous approach or accidental contact by persons or objects.

energized. Electrically connected to a source of potential difference, or electrically charged so as to have a potential significantly different from that of earth in the vicinity.

NOTE: **alive** or **live.**

equipment. A general term including fittings, devices, appliances, fixtures, apparatus, and similar terms used as part of or in connection with an electric supply or communication system.

explosion-proof apparatus. Apparatus enclosed in a case that is capable of withstanding an explosion of a specified gas or vapor which may occur within it and of preventing the ignition of a specified gas or vapor surrounding the enclosure by sparks, flashes, or explosion of the gas or vapor within, and which operates at such an external temperature that a surrounding flammable atmosphere will not be ignited thereby.

exposed. Not isolated or guarded.

fireproofing (of cables). The application of a fire-resistant covering.

grounded. Connected to or in contact with earth or connected to some extended conductive body which serves instead of the earth.

grounded effectively. *See:* **effectively grounded.**

grounded system. A system of conductors in which at least one conductor or point is intentionally grounded, either solidly or through a non-interrupting current-limiting device.

guarded. Covered, fenced, enclosed, or otherwise protected, by means of suitable covers or casings, barrier rails or screens, mats or platforms, designed to minimize the possibility, under normal conditions, of dangerous approach or accidental contact by persons or objects.

NOTE: Wires which are insulated, but not otherwise protected, are not considered as guarded.

handhole. An access opening, provided in equipment or in a below-the-surface enclosure in connection with underground lines, into which men reach but do not enter, for the purpose of installing, operating, or maintaining equipment or cable or both.

insulated. Separated from other conducting surfaces by a dielectric (including air space) offering a high resistance to the passage of current.

NOTE: When any object is said to be insulated, it is understood to be insulated for the conditions to which it is normally subjected. Otherwise, it is, within the purpose of these rules, uninsulated.

insulation (as applied to cable). That which is relied upon to insulate the conductor from other conductors or conducting parts or from ground.

insulation shielding. An envelope which encloses the insulation of a cable and provides an equipotential surface in contact with the cable insulation.

insulator. Insulating material in a form designed to support a conductor physically and electrically separate it from another conductor or object.

isolated. Not readily accessible to persons unless special means for access are used.

isolated by elevation. Elevated sufficiently so that persons may safely walk underneath.

isolator. *See:* disconnecting or isolating switch.

jacket. A protective covering over the insulation, core, or sheath of a cable.

joint use. Simultaneous use by two or more kinds of utilities.

lines.

1. **communication lines.** The conductors and their supporting or containing structures which are used for public or private signal or communication service, and which operate at potentials not exceeding 400 volts to ground or 750 volts between any two points of the circuit, and the transmitted power of which does not exceed 150 watts. When operating at less than 150 volts, no limit is placed on the transmitted power of the system. Under specified conditions, communication cables may include communication circuits exceeding the preceding limitation where such circuits are also used to supply power solely to communication equipment.

NOTE: Telephone, telegraph, railroad-signal, data, clock, fire, police-alarms, cable television and other systems conforming with the above are included. Lines used for signaling purposes, but not included under the above definition, are considered as supply lines of the same voltage and are to be so installed.

2. **electric supply lines.** Those conductors used to transmit electric energy and their necessary supporting or containing structures. Signal lines of more than 400 volts are always supply lines within the meaning of the rules, and those of less than 400 volts may be considered as supply lines, if so run and operated throughout. *Syn:* **supply lines.**

low voltage protection. The effect of a device operative on the reduction or failure of voltage so as to cause and maintain the interruption of power supply to the equipment protected.

manhole. A subsurface enclosure which personnel may enter and which is used for the purpose of installing, operating, and maintaining submersible equipment and cable.

manhole cover. A removable lid which closes the opening to a manhole or similar subsurface enclosure.

manhole grating. A grid which provides ventilation and a protective cover for a manhole opening.

manual. Capable of being operated by personal intervention.

pad mounted. A method of supporting equipment, generally at ground level.

prestressed concrete structures. Concrete structures which include metal tendons that are tensioned and anchored either before or after curing of the concrete.

pulling iron. An anchor secured in the wall, ceiling, or floor of a manhole or vault to attach rigging used to pull cable.

pulling tension. The longitudinal force exerted on a cable during installation.

qualified. Having adequate knowledge of the installation, construction or operation of apparatus and the hazards involved.

raceway. Any channel designed expressly and used solely for holding conductors.

random separation. Installed with no deliberate separation.

readily climbable. Having sufficient handholds and footholds to permit an average person to climb easily without using a ladder or other special equipment.

remotely operable (as applied to equipment). Capable of being operated from a position external to the structure in which it is installed or from a protected position within the structure.

roadway. The portion of highway, including shoulders, for vehicular use.

NOTE: A divided highway has two or more roadways. *See also:* **shoulder; traveled way.**

rural districts. All places not urban. This may include thinly settled areas within city limits.

sag.

1. The distance measured vertically from a conductor to the straight line joining its two points of support. Unless otherwise stated in the rule, the sag referred to is the sag at the midpoint of the span. See Fig D-1.

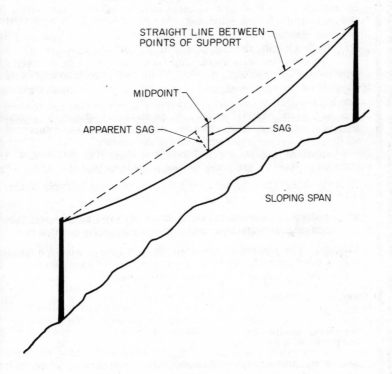

Fig D-1
Sag and Apparent Sag

2. **initial unloaded sag.** The sag of a conductor prior to the application of any external load.

3. **final sag.** The sag of a conductor under specified conditions of loading and temperature applied, after it has been subjected for an appreciable period to the loading prescribed for the loading district in which it is situated, or equivalent loading, and the loading removed. Final sag shall include the effect of inelastic deformation (creep).

4. **final unloaded sag.** The sag of a conductor after it has been subjected for an appreciable period to the loading prescribed for the loading district in which it is situated, or equivalent loading, and the loading removed. Final unloaded sag shall include the effect of inelastic deformation (creep).

5. **total sag.** The distance measured vertically from the conductor to the straight line joining its two points of support, under conditions of ice loading equivalent to the total resultant loading for the district in which it is located.

6. **maximum total sag.** The total sag at the midpoint of the straight line joining the two points of support of the conductor.

7. **apparent sag of a span.** The maximum distance between the wire in a given span and the straight line between the two points of support of the wire, measured perpendicularly from the straight line. See Fig D-1.

8. **sag of a conductor at any point in a span.** The distance measured vertically from the particular point in the conductor to a straight line between its two points of support.

9. **apparent sag at any point in the span.** The distance, at the particular point in the span, between the wire and the straight line between the two points of support of the wire, measured perpendicularly from the straight line.

service drop. The overhead conductors between the electric supply or communication line and the building or structure being served.

shoulder. The portion of the roadway contiguous with the traveled way for accommodation of stopped vehicles for emergency use and for lateral support of base and surface course.

side-wall pressure. The crushing force exerted on a cable during installation.

span length. The horizontal distance between two adjacent supporting points of a conductor.

span wire. An auxiliary suspension wire which serves to support one or more trolley contact conductors or a light fixture and the conductors which connect it to a supply system.

structure conflict. A line is so situated with respect to a second line that the overturning of the first line will result in contact between its supporting structures or conductors and the conductors of the second line, assuming that no conductors are broken in either line.

supply equipment. *See:* electric supply equipment.

supply station. *See:* electric supply station.

supporting structure. The main supporting unit (usually a pole or tower).

susceptiveness. The characteristics of a communications circuit including its connected apparatus which determine the extent to which it is adversely affected by inductive fields.

switch. A device for opening and closing or for changing the connection of a circuit. In these rules, a switch is understood to be manually operable, unless otherwise stated.

switchboard. A type of switchgear assembly that consists of one or more panels with electric devices mounted thereon, and associated framework.

system operator. A person designated to operate the system or parts thereof.

tag. Accident prevention tag (DANGER, MEN AT WORK, etc) of a distinctive appearance used for the purpose of personnel protection to indicate that the operation of the device to which it is attached is restricted.

tension, unloaded.
1. **initial.** The longitudinal tension in a conductor prior to the application of any external load.
2. **final.** The longitudinal tension in a conductor after it has been subjected for an appreciable period to the loading prescribed for the loading district in which it is situated, or equivalent loading, and the loading removed. Final unloaded tension shall include the effect of inelastic deformation (creep).

termination. *See:* cable terminal.

transformer vault. An isolated enclosure either above or below ground with fire-resistant walls, ceiling, and floor, in which transformers and related equipment are installed, and which is not continuously attended during operation. *See also:* vault.

traveled way. The portion of the roadway for the movement of vehicles, exclusive of shoulders and full-time parking lanes.

urban districts. Thickly settled areas (whether in cities or suburbs) or where congested traffic often occurs. A highway, even though in thinly settled areas, on which the traffic is often very heavy, is considered as urban.

utility. An organization responsible for the installation, operation or maintenance of electric supply or communication systems.

utilization equipment. Equipment, devices, and connected wiring which utilize electric energy for mechanical, chemical, heating, lighting, testing, or similar purposes and are not a part of supply equipment, supply lines, or communication lines.

vault. An enclosure above or below ground which personnel may enter and is used for the purpose of installing, operating, or maintaining equipment or cable which need not be of a submersible design.

voltage.

1. The effective (rms) potential difference between any two conductors or between a conductor and ground. Voltages are expressed in nominal values unless otherwise indicated. The nominal voltage of a system or circuit is the value assigned to a system or circuit of a given voltage class for the purpose of convenient designation. The operating voltage of the system may vary above or below this value.

2. **voltage of circuit not effectively grounded.** The highest nominal voltage available between any two conductors of the circuit.

NOTE: If one circuit is directly connected to and supplied from another circuit of higher voltage (as in the case of an autotransformer), both are considered as of the higher voltage, unless the circuit of the lower voltage is effectively grounded, in which case its voltage is not determined by the circuit of higher voltage. Direct connection implies electric connection as distinguished from connection merely through electromagnetic or electrostatic induction.

3. **voltage of a constant current circuit.** The highest normal full load voltage of the current.

4. **voltage of an effectively grounded circuit.** The highest nominal voltage available between any conductor of the circuit and ground unless otherwise indicated.

5. **voltage to ground of:**

    a. **a grounded circuit.** The highest nominal voltage available between any conductor of the circuit and that point or conductor of the circuit which is grounded.

    b. **an ungrounded circuit.** The highest nominal voltage available between any two conductors of the circuit concerned.

6. **voltage to ground of a conductor of:**

a. **a grounded circuit.** The nominal voltage between such conductor and that point or conductor of the circuit which is grounded.

b. **an ungrounded circuit.** The highest nominal voltage between such conductor and any other conductor of the circuit concerned.

wire gages. Throughout these rules the American Wire Gage (AWG), formerly known as Brown & Sharpe (B&S), is the standard gage for copper, aluminum and other conductors, excepting only steel conductors for which the Steel Wire Gage (Stl WG) is used.

NOTE: The Birmingham Wire Gage is obsolete.

Section 9 of
ANSI
C2-1981
Revision of
Section 9 of
C2.2-1976

G

American National Standard

National Electrical Safety Code

Section 9. Grounding Methods for Electric Supply and Communication Facilities

Section 9. Grounding Methods for Electric
Supply and Communication Facilities

90.   Purpose

The purpose of Section 9 of this code is to provide practical
methods of grounding, as one of the means of safeguarding
employees and the public from injury that may be caused by
electrical potential.

91.   Scope

Section 9 of this code covers methods of protective
grounding of supply and communication conductors and
equipment. The rules requiring grounding are in other parts
of this code.

These rules do not cover the grounded return of electric
railways nor those lightning protection wires which are
normally independent of supply or communication wires or
equipment.

92.   Point of Connection of Grounding Conductor

A. Direct Current Systems Which Are To Be Grounded
1. 750 volts and below
    Connection shall be made only at supply stations. In
    three-wire direct-current systems the connection shall be
    made to the neutral.
2. Over 750 volts
    Connection shall be made at both the supply and load
    stations. The connection shall be made to the neutral of
    the system. The ground or grounding electrode may be
    external to or remotely located from each of the
    stations.

    One of the two stations may have its ground connection
    made through surge arresters provided the other station
    neutral is effectively grounded as described above.

B. Alternating Current Systems Which Are To Be
    Grounded
1. 750 volts and below
    The point of the grounding connection on a wye-con-
    nected three-phase four-wire system, or on a single-phase
    three-wire system, shall be the neutral conductor. On

other one-, two-, or three-phase systems with an associated lighting circuit or circuits, the point of grounding connection shall be on the common circuit conductor associated with the lighting circuits.

The point of grounding connection on three-phase three-wire system, whether derived from a delta connected or an ungrounded wye connected transformer installation not used for lightning, may be any of the circuit conductors, or it may be a separately derived neutral.

The grounding connections shall be made at the source, and at the line side of all service equipment.

2. Over 750 volts
   a. Nonshielded (Bare or Covered Conductors or Insulated Non-Shielded Cables).

   Grounding connection shall be made at the neutral of the source. Additional connections may be made, if desired, along the length of the neutral, where this is one of the system conductors.

   b. Shielded
      (1) Surge-Arrester Cable-Shielding Interconnection. Cable shielding grounds shall be bonded to surge arrester grounds, where provided, at points where underground cables are connected to overhead lines.
      (2) Cable Without Insulating Jacket

      Connection shall be made to the neutral of the source transformer and at cable termination points.
      (3) Cable With Insulating Jacket

      Additional bonding and connections between the cable insulation shielding or sheaths and the system ground are recommended. In multi-grounded shielded cable systems, the shielding (including sheath) shall be grounded at each cable joint exposed to personnel contact. Where multi-grounded shielding cannot be used for electrolysis or sheath current reasons, the shielding sheaths and splice enclosure devices shall be insulated for the voltage which may appear on them during normal operation.

      Bonding transformers or reactors may be substituted for direct ground connection at one end of the cable.

3. Separate Grounding Conductor

If a separate grounding conductor is used as an adjunct to a cable run underground, it shall be connected at the source transformer and at cable accessories where these are to be grounded. This grounding conductor shall be located in the same direct burial or duct bank run (or the same duct if this is of magnetic material) as the circuit conductors.

EXCEPTION: The grounding conductor for a circuit which is installed in a magnetic duct need not be in the same duct if the duct containing the circuit is bonded to the separate grounding conductor at both ends.

C. Messenger Wires and Guys

1. Messenger Wires

Messenger wires required to be grounded shall be connected to grounding conductors at poles or structures at maximum intervals as listed below.

a. Where messenger wires are adequate for system grounding conductors (Rules 93C1, 93C2, and 93C5), four connections in each mile.

b. Where messenger wires are not adequate for system grounding conductors, eight connections per mile, exclusive of service grounds.

2. Guys

Guys which are required to be grounded shall be connected to:

a. Grounded steel structures or to an effective ground connection on wood poles.

b. A line conductor which has at least four ground connections in each mile of line in addition to the ground connections at individual services.

D. Current in Grounding Conductor

Ground connection points shall be so arranged that under normal circumstances there will be no objectionable flow of current over the grounding conductor. If an objectionable flow of current occurs over a grounding conductor due to the use of multiple grounds, one or more of the following should be used:

1. Abandon one or more grounds.
2. Change location of grounds.
3. Interrupt the continuity of the conductor between ground connections.

4.   Subject to the approval of the administrative authority take other effective means to limit the current.

The system ground of the source transformer shall not be removed.

The temporary currents set up under abnormal conditions while the grounding conductors are performing their intended protective functions are not considered objectionable. The conductor shall have the capability of conducting anticipated fault current without thermal overloading or excessive voltage buildup. Refer to Rule 93C.

E.   Fences

Fences, where required to be grounded by other parts of this code, shall be grounded at or near the location of a supply line or lines crossing them, and additionally, at distances not exceeding 150 ft on either side. Fences shall also be grounded at each side of a gate or other opening in the fence. Any gate or other opening shall also be bonded across by a buried bonding jumper. A gate shall be metallically connected or bonded to the grounding conductor, jumper, or fence. Separate barbed wire strands above fencing, on nonconducting posts, shall be bonded to metallic fencing or grounding conductors at the grounding points.

Where required to be grounded, fences shall be bonded to the grounding system of the enclosed equipment or to a separate underground conductor below or near the fence line.

93.   Grounding Conductor and Means of Connection

A.   Composition of Grounding Conductors

In all cases the grounding conductor shall be made of copper or other metals or combinations of metals which will not corrode excessively during the expected service life under the existing conditions and, if practical, shall be without joint or splice. If joints are unavoidable, they shall be so made and maintained as to not materially increase the resistance of the grounding conductor and shall have appropriate mechanical and corrosion resistant characteristics. For surge arresters and ground detectors, the grounding conductor or conductors shall be as short, straight, and free from sharp bends as practical. The structural metal frame of a building or structure may serve as a grounding conductor to an acceptable grounding electrode.

In no case shall a circuit-opening device be inserted in the grounding conductor or connection except where its operation will result in the automatic disconnection from all sources of energy of the circuit leads connected to the equipment so grounded.

EXCEPTION: Temporary disconnection of grounding conductors for testing purposes, under competent supervision, shall be permitted.

B. Connection of Grounding Conductors

Connection of the grounding conductor shall be made by a means matching the characteristics of both the grounded and grounding conductors, and suitable for the environmental exposure. These means include brazing, welding, mechanical and compression connections, ground clamps, and ground straps. Soldering is acceptable only in conjunction with lead sheaths.

C. Ampacity and Strength

The "short time ampacity" of a bare grounding conductor is that current which the conductor can carry for the time during which the current flows without melting or separating under the applied tensions. If a grounding conductor is insulated, its short time ampacity is the current which it can carry for the applicable time without damaging the insulation. Where grounding conductors at one location are paralleled, the increased total current capacity may be considered.

1. System Grounding Conductors for Single-Grounded Systems
    The system grounding conductor or conductors for a system with single system grounding electrode or set of electrodes, exclusive of grounds at individual services, shall have a short time ampacity adequate for the fault current which can flow in the grounding conductor or conductors for the operating time of the system protective device. If this value cannot be readily determined, continuous ampacity of the grounding conductor or conductors shall be not less than the full load continuous current of the system supply transformer or other source of supply.

2. System Grounding Conductors for Multigrounded Alternating Current Systems
    The system grounding conductors for an alternating current system with grounds at more than one location exclusive of grounds at individual services shall have con-

tinuous total ampacities at each location of not less than one-fifth that of the conductors to which they are attached. (See also Rule 93C8).

3. Grounding Conductors for Instrument Transformers
The grounding conductor for instrument cases and secondary circuits of instrument transformers shall not be smaller than No 12 AWG copper or have equivalent ampacity.

4. Grounding Conductors for Primary Surge Arresters
The grounding conductor or conductors shall have adequate short time ampacity under conditions of excess current caused by or following a surge. Individual arrester grounding conductors shall be no smaller than No 6 AWG copper or No 4 AWG aluminum.

EXCEPTION: Arrester grounding conductors may be copper-clad or aluminum-clad steel wire having not less than 30 percent of the conductivity of solid copper or aluminum wire of the same diameter.

Where flexibility of the grounding conductor, such as adjacent to the base of the arrester, is vital to its proper operation, a suitably flexible conductor shall be employed.

5. Grounding Conductors for Equipment, Messenger Wires, and Guys
   a. Conductors
   The grounding conductors for equipment, raceways, cable, messenger wires, guys, sheaths, and other metal enclosures for wires shall have short time ampacities adequate for the available fault current and operating time of the system fault protective device. If no overcurrent or fault protection is provided, the ampacity of the grounding conductor shall be determined by the design and operating conditions of the circuit, but shall not be less than that of No 8 AWG copper. Where the adequacy and continuity of the conductor enclosures and their attachment to the equipment enclosures is assured, this path can constitute the equipment grounding conductor.
   b. Connections
   Connection of the grounding conductor shall be to a suitable lug, terminal, or device not disturbed in normal inspection, maintenance, or operation.

6. Fences
The grounding conductor for fences required to be

grounded by other parts of this code shall be any of those meeting the requirements of Rule 93C5 or shall be steel wire not smaller than No 5 Steel Wire Gage. It shall be connected to the fence posts with connecting means suitable for the material when the posts are of conducting material. If the posts are of nonconducting material, suitable bonding connections shall be made to the fence mesh strands and the barbed wire strands at each grounding conductor point.

7. Bonding of Equipment Frames and Enclosures
Where required, a low impedance metallic path shall be provided for the passage of possible conductor or equipment, or both, fault current back to the grounded terminal of the supply, where the supply is local. Where the supply is remote, the metallic path shall interconnect the equipment frames and enclosures with all other nonenergized conducting components within reach and shall additionally be connected to ground as outlined in Rule 93C5. Short-time ampacities of bonding conductors shall be adequate for the duty involved.

8. Ampacity Limit
No grounding conductor need have greater ampacity than either:
   a. The phase conductors which would supply the ground fault current, or
   b. The maximum current which can flow through it to the ground electrode or electrodes to which it is attached. For a single grounding conductor and connected electrode or electrodes, this would be the supply voltage divided by the electrode resistance (approximately).

9. Strength
All grounding conductors shall have mechanical strength suitable for the conditions to which they may reasonably be subjected.

Further, unguarded grounding conductors shall have a tensile strength not less than that of No 8 AWG softdrawn copper, except as noted in Rule 93C3.

D. Guarding and Protection
1. The grounding conductors for single grounded systems and those exposed to mechanical damage shall be guarded. However, grounding conductors need not be guarded where not readily accessible to the public nor where grounding multigrounded circuits or equipment.

2. Where guarding is required, grounding conductors shall be protected by guards suitable for the exposure to which they may reasonably be subjected. The guards should extend for not less than 8 ft above the ground or platform from which the grounding conductors are accessible to the public.

3. Where guarding is not required, grounds shall be protected by being substantially attached closely to the surface of the pole or other structure in areas of exposure to mechanical damage and, where practical, on the portion of the structure having least exposure.

4. Guards used for grounding conductors of lightning protection equipment shall be of nonmagnetic materials if the guard completely encloses the grounding conductor or is not bonded at both ends to the grounding conductor.

E. Underground

1. Grounding conductors laid directly underground shall be laid slack or shall be of sufficient strength to prevent being readily broken by earth movement or settling normal at the particular location.

2. Direct-buried uninsulated joints or splices in grounding conductors should be welded, brazed, or of the compression type to minimize the possibility of loosening or corrosion. The number of joints or splices should be the minimum practical.

3. Grounding cable insulation shielding systems shall be interconnected with all other accessible grounded power supply equipment in manholes, handholes, and vaults.
   EXCEPTION: Where cathodic protection or shield cross-bonding is involved, interconnection may be omitted.

4. Looped magnetic elements such as structural steel, piping, reinforcing bars, etc, should not separate grounding conductors from the phase conductors of circuits they serve.

5. Metals used for grounding, in direct contact with earth, concrete, or masonry, shall have been proven suitable for such exposure.

   NOTE 1: Under present technology, aluminum has not generally been proven suitable for such use.
   NOTE 2: Metals of different galvanic potentials which are electrically interconnected may require protection against galvanic corrosion.

6. Sheath Transposition Connections (Cross-Bonding)

a. Where cable insulating shields or sheaths, which are normally connected to ground, are insulated from ground to minimize shield circulating currents, they shall be insulated from personnel contact at accessible locations. Transposition connections and bonding jumpers shall be insulated for nominal 600 volt service, unless the normal shielding voltage exceeds this level, in which case the insulation shall be ample for the working voltage to ground.

b. Bonding jumpers and connecting means shall be sized and selected to carry the available fault current without damaging jumper insulation or sheath connections.

F. Common Grounding Conductor for Circuits, Metal Raceways, and Equipment

Where the ampacity of a supply system grounding conductor is also adequate for equipment grounding requirements, this conductor may be used for the combined purpose. Equipment referred to includes the frames and enclosures of supply system control and auxiliary components, conductor raceways, cable shields, and other enclosures.

94. Grounding Electrodes

The grounding electrode shall be permanent and adequate for the electrical system involved. A common electrode or electrode system shall be employed for grounding the electrical system and the conductor enclosures and equipment served by that system. This may be accomplished by interconnecting these elements at the "Point of Connection of Grounding Conductor," Rule 92.

Grounding electrodes shall be one of the following:

A. Existing Electrodes

Existing electrodes consist of conducting items installed for purposes other than grounding:

1. Metallic Water Piping System

Extensive metallic underground cold water piping systems may be used as grounding electrodes.

NOTE: Such systems normally have very low resistance to earth and have been extensively used in the past. They are the preferred electrode type where they are readily accessible.

EXCEPTION: Water systems with nonmetallic noncurrent-carrying pipe or insulating joints are not suitable for use as grounding electrodes.

2. Local Systems
Isolated buried metallic cold water piping connecting to wells having sufficiently low measured resistance to earth may be used as grounding electrodes.

NOTE: Care should be exercised to insure that all parts that might become disconnected are effectively bonded together.

3. Steel Reinforcing Bars in Concrete Foundations and Footings
The reinforcing bar system of a concrete foundation or footing which is not insulated from direct contact with earth, and which extends at least 3 feet below grade, constitutes an effective and acceptable type of grounding electrode. Where steel supported on this foundation is to be used as a grounding conductor (tower, structure, etc), it shall be interconnected by bonding between anchor bolts and reinforcing bars or by cable from the reinforcing bars to the structure above the concrete.
The normally applied steel ties are considered to provide adequate bonding between bars of the reinforcing cage.

NOTE: Where reinforcing bars in concrete are not suitably connected to a metal structure above the concrete, and the latter structure is subjected to grounding discharge currents (even connected to another electrode), there is likelihood of damage to the intervening concrete from ground-seeking current passing through the semi-conducting concrete.

B. Made Electrodes
1. General
Where made electrodes are used, they shall as far as practical penetrate into permanent moisture level and below the frostline. Made electrodes shall be of metal or combinations of metals which do not corrode excessively under the existing conditions for the expected service life.
All outer surfaces of made electrodes shall be conductive, that is, not having paint, enamel, or other insulating type covering.

2. Driven Rods
Driven rods may be sectional; the total length shall not

be less than 8 feet. Driven depth shall be 8 feet minimum. The upper end shall be flush with or below the ground level unless suitably protected. Longer rods or multiple rods may be used to reduce the ground resistance. Spacing between multiple rods should not be less than 6 feet.

EXCEPTION: Where rock bottom is encountered, driven depth may be less than 8 feet, or other types of electrode employed.

Iron or steel rods shall have minimum cross-sectional dimension of ⅝ inch. Copper-clad, stainless steel, or stainless steel-clad rods shall have a minimum cross-sectional dimension of ½ inch.

3. Buried Wire, Strips, or Plates
    In areas of high soil resistivity or shallow bedrock, or where lower resistance is required than attainable with driven rods, one or more of the following electrodes may be more useful:

    a. Wire
        Bare wires 0.162 inch in diameter or larger, conforming to Rule 93E5, buried in earth at a depth not less than 18 inches and not less than 100 feet total in length, laid approximately straight, constitutes an acceptable made electrode. (This is frequently designated a "counterpoise.") The wire may be in a single length, or may be several lengths connected together at ends or at some point away from the ends.

        The wire may take the form of a network with many parallel wires spaced in two-dimensional array, referred to as a grid.

        EXCEPTION 1: Where rock bottom is encountered, burial depth may be less than 18 inches.

        EXCEPTION 2: Other lengths or configurations may be used if their suitability is supported by a qualified engineering study.

    b. Strips
        Strips of metal not less than 10 feet in total length and with total (two side) surface not less than 5 square feet buried in soil at a depth not less than 18 inches constitutes an acceptable made electrode. Ferrous metal electrodes shall be not less than ¼ inch in thickness and nonferrous metal electrodes not less than 0.06 inches.

NOTE: Strip electrodes are frequently useful in rocky areas where only irregularly shaped pits are practical to excavate.

c. Plates or Sheets

Metal plates or sheets having not less than 2 square feet of surface exposed to the soil, and at a depth of not less than 5 feet, constitute an acceptable made electrode. Ferrous metal electrodes shall be not less than $\frac{1}{4}$ inch in thickness and nonferrous metal electrodes not less than 0.06 inches.

4. Pole Butt Plates and Wire Wraps

a. General

In areas of very low soil resistivity there are two constructions, described in specifications b and c below, which may provide effective grounding electrode functions although they are inadequate in most other locations. Where these have been proven to have adequately low earth resistance by the application of Rule 96, two such electrodes may be counted as one made electrode and ground for application of Rules 92C1a, 92C2b, 97C, and 96A3; however, these types shall not be the sole grounding electrode at transformer locations.

b. Pole Butt Plates

Subject to the limitations of Rule 94B4a, a pole butt plate on the base of a wooden pole, possibly folded up around the base of the pole butt, may be considered an acceptable electrode in locations where the limitations of Rule 96 are met. The plates shall not be less than $\frac{1}{4}$ inch thick if of ferrous metal, and not less than 0.06 inch thick if of non-ferrous metal. Further, the minimum plate area exposed to the soil shall be 0.5 sq ft.

c. Wire Wrap

Subject to the limitations of Rule 94B4a, made electrodes may be wire attached to the pole previous to the setting of the pole. The wire shall be of copper or other metals which will not corrode excessively under the existing conditions and shall have a continuous bare or exposed length below ground level of not less than 12 ft, shall extend to the bottom of the pole, and shall not be smaller than No 6 AWG.

5. Concentric Neutral Cable

Systems employing extensive (100 ft minimum length) buried bare concentric neutral cable in contact with the earth may employ the concentric neutral as a grounding electrode. The concentric neutral may be covered with a semi-conducting jacket which has a radial resistivity not exceeding 20 meter ohms and which will remain essentially stable in service. The radial resistivity of the jacket material is that value calculated from measurements on a unit length of cable, of the resistance between the concentric neutral and a surrounding conducting medium. Radial resistivity equals resistance of unit length times the surface area of jacket divided by the average thickness of the jacket over the neutral conductors. All dimensions are to be expressed in meters.

6. Concrete-Encased Electrodes

A metallic wire, rod, or structural shape, meeting Rule 93E5 and encased in concrete which is not insulated from direct contact with earth shall constitute an acceptable ground electrode. The concrete depth below grade shall not be less than 1 foot, and a depth of 2½ feet is recommended. Wire shall be no smaller than No. 4 AWG if copper, or ⅜ inch diameter if steel. It shall be not less than 20 feet long, and shall remain entirely within the concrete except for the external connection. The conductor should be run as straight as practical.

The metal elements may be composed of a number of shorter lengths arrayed within the concrete and connected together (for example, the reinforcing system in a structural footing).

EXCEPTION: Other wire length or configurations may be used if their suitability is supported by a qualified engineering study.

NOTE1: The lowest resistance per unit wire length will result from a straight wire installation.
NOTE 2: The outline of the concrete need not be regular, but may conform to an irregular or rocky excavation.
NOTE 3: Concrete encased electrodes are frequently more practical or effective than driven rods or strips or plates buried directly in earth.

95.   Method of Connection to Electrode
A.   Ground Connections

The ground connections shall be as accessible as practical and shall be made to the electrode by methods providing the required permanence and ampacity, such as:

1. A permanently effective clamp, fitting, braze, or weld.
2. A bronze plug which has been tightly screwed into the electrode.
3. For steel-framed structures employing a concrete-encased reinforcing bar electrode, a steel rod similar to the reinforcing bar shall be used to join, by welding, a main vertical reinforcing bar to an anchor bolt. The bolt shall be substantially and permanently connected to the baseplate of the steel column supported on that footing. The electrical system may then be connected (for grounding) to the building frame by welding or by a bronze bolt tapped into a structural member of that frame.
4. For nonsteel frame structures employing a concrete-encased rod or wire electrode, an insulated copper conductor of size meeting the requirements of Rule 93C (except not smaller than No. 4 AWG) shall be connected to the steel rod or wire using a cable clamp suitable for steel cable. This clamp and all the bared portion of the copper conductor including ends of exposed strands within the concrete shall be completely covered with mastic or sealing compound before concrete is poured to minimize the possibility of galvanic corrosion. The copper conductor end shall be brought to or out of the concrete surface at the required location for connection to the electrical system. If the copper wire is carried beyond the surface of the concrete, it shall be no smaller than No 2 AWG.

   Alternatively, the copper wire may be brought out of the concrete at the bottom of the hole and carried external to the concrete for surface connection.

B.   Point of Connection to Piping Systems

1. The point of connection of a grounding conductor to a metallic water piping system shall be as near as is practical to the water-service entrance to the building or near the equipment to be grounded and shall be accessible. If a water meter is between the point of connection and the underground water pipe, the metallic water piping system shall be made electrically continuous by bonding together all parts between the connection and the pipe

entrance which may become disconnected, such as meters and service unions.

2. Made grounds or grounded structures should be separated by 10 feet or more from pipelines used for the transmission of flammable liquids or gases operating at high pressures (150 pounds per square inch or greater) unless they are electrically interconnected and cathodically protected as a single unit. Grounds within 10 feet of such pipelines should be avoided or shall be coordinated so that hazardous alternating current conditions will not exist and cathodic protection of the pipeline will not be nullified.

C. Contact Surfaces

If any coating of nonconducting material, such as enamel, rust, or scale, is present on electrode contact surfaces, at the point of connection, such a coating shall be thoroughly removed where required to obtain the requisite good connection. Special fittings so designed as to make such removal of nonconducting coatings unnecessary may also be used.

96. Ground Resistance

A. Requirements

The grounding electrode system may consist of one or more interconnected electrodes. It shall have a resistance to ground low enough to minimize hazards to personnel and to permit prompt operation of circuit protective devices.

1. Supply Stations

Where very high voltages and currents are involved, such as in large substations, extensive grounding grid systems of multiple buried wires and rods and other protective means may be required.

NOTE: It is recommended that the combination of maximum local ground fault current and impedance of the grounding system not exceed values which will limit exposure potentials to the following:

$$E_{\text{step}} = (1000 + 6\rho s) \frac{0.116}{\sqrt{t}}$$

$$E_{\text{touch}} = (1000 + 1.5\rho s) \frac{0.116}{\sqrt{t}}$$

where

E_{step} maximum tolerable voltage difference between any two points on the ground surface which can be touched simultaneously by 2 (separated) feet

E_{touch} maximum tolerable voltage difference between any point on the ground where a man may stand and any point which can be touched simultaneously by either hand

ρs resistivity of the soil near the surface in ohm-meters (divide the ohm-centimeter value by 100 to obtain this)

t time of exposure in seconds (clearing time of system overcurrent equipment)

2. Single Grounded (Unigrounded or Delta) Systems.

Individual made electrodes shall, where practical, have a resistance to ground not exceeding 25 ohms. If a single electrode resistance exceeds 25 ohms, two electrodes connected in parallel shall be used.

3. Multiple Grounded Systems.

The neutral, which shall be of sufficient size and ampacity for the duty involved, shall be connected to made electrodes at each transformer location and at a sufficient number of additional points to total not less than four grounds in each mile of line, not including grounds at individual services:

NOTE: Multiple grounding systems extending over a substantial distance are more dependent on the multiplicity of grounding electrodes than on the resistance to ground of any individual electrode. Therefore, no specific values are imposed for the resistance of individual electrodes.

97.   Separation of Grounding Conductors

A. Except as permitted in Rule 97B, grounding conductors from equipment and circuits of each of the following classes shall be run separately to the grounding electrode for each of the following classes:

1. Surge arresters of circuits over 750 volts, and frames of any equipment operating at over 750 volts.
2. Lighting and power circuits under 750 volts.
3. Lightning rods, unless attached to a grounded metal supporting structure.

Alternatively, the grounding conductors shall be run

separately to a sufficiently heavy ground bus or system ground cable which is well connected to ground at more than one place.

B. The grounding conductors of either of the equipment classes detailed in Rules 97A1 and 97A2 may be interconnected. utilizing a single grounding conductor, provided:

1. There is a direct earth grounding connection at each arrester location.

2. The secondary neutral is common with, or connected to, a primary neutral meeting the grounding requirements of Rule 97C.

C. Primary and secondary circuits utilizing a single conductor as a common neutral shall have at least four ground connections on such conductor in each mile of line, exclusive of ground connections at customers' service equipment.

D. Where the secondary neutral is not interconnected with the primary neutral as in Rule 97 B, interconnection of the neutral may be made through a spark gap. The gap shall have a 60 hertz breakdown voltage of at least twice the primary circuit voltage but not necessarily more than 10 kilovolts. At least one other ground connection on the secondary neutral shall be provided, at a distance of not less than 20 feet from the surge arrester grounding electrode.

E. Where separate electrodes are used for system isolation, separate grounding conductors shall be used. Where multiple electrodes are used to reduce grounding resistance, they may be bonded together and connected to a single grounding conductor.

F. Made electrodes used for grounding surge arresters of ungrounded supply systems operated at potentials exceeding 15 kilovolts phase to phase should be located at least 20 feet from buried communications cables. Where lines with lesser separations are to be constructed, reasonable advance notice should be given to the owners or operators of the affected systems.

98.   Rule 98 not used in this edition.

99.   **Grounding Methods for Telephone and Other Communication Apparatus on Circuits Exposed to Supply Lines or Lightning**
Protectors and, where required, exposed noncurrent-carrying metal parts located in central offices or outside installations shall be grounded in the following manner.

A. Electrode

The grounding conductor shall be connected to an acceptable grounding electrode as described in Rule 94, with the following additions and exception:

1. Connection may be made to the metallic supply, service conduit, service-equipment enclosure, or grounding electrode conductor where the grounded conductor of the supply service is connected to an acceptable water pipe electrode at the building.

2. Where the grounding means in Rules 94A1 or 99A1 are not available, the grounding conductor shall be connected to the metallic supply service conduit, service-equipment enclosure, grounding electrode conductor, or grounding electrode of the supply service of a multi-grounded neutral power supply.

EXCEPTION: A variance to Rule 94B2 is allowed for this application. Iron or steel rods may have a minimum cross-sectional dimension of ½ inch and a length of 5 feet. The driven depth shall be 5 feet minimum, subject to the exception of Rule 94B2.

B. Electrode Connection

The grounding conductor shall preferably be made of copper (or other material which will not corrode excessively under the prevailing conditions of use) and shall be not less than No. 14 AWG (0.064 inch) in size. The grounding conductor shall be attached to the electrode by means of a bolted clamp or other suitable methods.

C. Bonding of Electrodes

A bond not smaller than No. 6 AWG (0.162 inch) copper or equivalent shall be placed between the communication grounding electrode and the supply system neutral grounding electrode where separate electrodes are used in or on the same building or structure.

American National Standard

National Electrical Safety Code

Part 1 (Sections 10-19). Rules for the Installation and Maintenance of Electric Supply Stations and Equipment

Supersedes
National Bureau of Standards Handbook 110-1
C2.1-1971

PART 1. Rules for the Installation and Maintenance of Electric Supply Stations and Equipment

Section 10. Purpose and Scope of Rules

100. Purpose

The purpose of Part 1 of this code is the practical safeguarding of persons during the installation, operation, or maintenance of electric supply stations and their associated equipment.

101. Scope

Part 1 of this code covers the electric supply conductors and equipment, along with the associated structural arrangements in electric supply stations, which are accessible only to qualified personnel. It also covers the conductors and equipment employed primarily for the utilization of electric power when such conductors and equipment are used by the utility in the exercise of its function as a utility.

Section 11. Protective Arrangements in Electric Supply Stations

110. General Requirements

A. Enclosure of Equipment

Rooms and spaces in which electric supply conductors or equipment are installed shall be so arranged with fences, screens, partitions or walls as to minimize the possibility of entrance of unauthorized persons or interference by them with equipment inside. Entrances not under observation of an authorized attendant shall be kept locked.

Warning signs shall be displayed at entrances.

Metal fences when used to enclose electric supply stations having energized electrical conductors or equipment that can be reached by trespassers, shall be a minimum of seven feet in height and shall be effectively grounded. Other types of construction such as nonmetallic material shall present equivalent barriers to climbing or other unauthorized entry.

NOTE: It is recommended that, where permissable a one foot extension, carrying three strands of barbed wire, be used above the fence fabric, either as an outside or inside the fence overhang, or as a vertical extension of the fence to obtain the desired overall height.

B. Rooms and Spaces

All rooms and spaces in which electric supply equipment is installed shall comply with the following requirements.

1. Construction

They shall be as much as practical noncombustible.

2. Use

They should be as much as practical free from combustible materials, dust, and fumes and shall not be used for manufacturing or for storage, except for minor parts essential to the maintenance of the installed equipment. (For battery areas, see Section 14; for auxiliary equipment in hazardous locations, see Rule 127).

3. Ventilation

There should be sufficient ventilation to maintain operating temperatures within ratings, arranged to minimize accumulation of airborne contaminants under any operating conditions.

4. Moisture and Weather

They should be dry. In outdoor stations or stations in wet tunnels, subways or other moist or high humidity locations, the equipment shall be suitably designed to withstand the prevailing atmospheric conditions.

C. Electric Equipment

All stationary equipment shall be supported and secured in place.

111. Illumination

A. Under Normal Conditions

Rooms and spaces shall have means for artificial illumination. The illumination levels listed in Table 110-1 are recommended minimum footcandles for safety to be maintained on the task.

Table 111-1. Illumination Levels

Location	Recommended Minimum Footcandles
Central Station	
Air conditioning equipment, air preheater and fan floor, ash sluicing	5
Auxiliaries, battery areas, boiler feed pumps, tanks, compressors, gage area	10
Boiler Platforms	5
Burner Platforms	10
Cable Room, circulator, or pump bay	5
Chemical Laboratory	25
Coal conveyor, crusher, feeder, scale areas, pulverizer, fan area, transfer tower	5
Condensers, deaerator floor, evaporator floor, heater floors	5
Control Rooms	
Vertical face of switchboards	
Simples or section of duplex operator:	
Type A — Large centralized control room 66 inches above floor	25
Type B — Ordinary control room 66 inches above floor	15
Section of duplex facing away from operator	15
Bench boards (horizontal level)	25
Area inside duplex switchboards	5
Rear of all switchboard panels (vertical)	5
Dispatch boards	
Horizontal plane (desk level)	25
Vertical face of board (48 inches above floor, facing operator):	
System load dispatch room	25
Secondary dispatch room	15
Hydrogen and carbon dioxide manifold area	10
Precipitators	5
Screen House	10

(Continued on page 88)

Table 111-1   *(Continued)*

Soot or slag blower platform	5
Steam headers and throttles	5
Switchgear, power	10
Telephone equipment room	10
Tunnels or galleries, piping	5
Turbine bay sub-basement	10
Turbine Room	15
Visitor's gallery	10
Water Treating Area	10

Central Station (Exterior)

Catwalks	2
Cinder dumps	0.2
Coal Storage Area	0.2
Coal unloading	
Dock (loading or unloading zone)	5
Barge storage area	0.5
Car dumper	0.5
Tipple	5
Conveyors	2
Entrances	
Generating or Service Building	
Main	10
Secondary	2
Gate House	
Pedestrian Entrance	10
Conveyor entrance	5
Fence	0.2
Fuel-oil delivery headers	5
Oil storage tanks	1
Open yard	0.2
Platforms — boiler, turbine deck	5
Roadway	
Between or along buildings	1
Not bordered by buildings	0.5
Substation	
General horizontal	2
Specific vertical (on disconnects)	2

B. **Emergency Lighting**
   1. A separate emergency source of illumination with automatic initiation, from an independent generator, storage battery or other suitable source, shall be provided in every attended station.
   2. Emergency lighting of one footcandle shall be provided in exit paths from all areas of attended stations. Consideration must be given to the type of service to be rendered whether of short time or long duration. The minimum duration shall be one and one-half hours. It is recommended that emergency circuit wiring shall be kept independent of all other wiring and equipment.

C. **Fixtures**
   Arrangements for permanent fixtures and plug receptacles shall be such that portable cords need not be brought into dangerous proximity to live or moving parts. All lighting shall be controlled and serviced from safely accessible locations.

D. **Attachment Plugs and Receptacles for General Use**
   Portable conductors shall be attached to fixed wiring only through separable attachment plugs which will disconnect all poles by one operation. Receptacles installed on two or three wire single phase, ac branch circuits shall be of the grounding type. Receptacles connected to circuits having different voltages, frequencies or types of current (ac or dc) on the same premises shall be of such design that attachment plugs used on such circuits are not interchangeable.

E. **Receptacles in Damp or Wet Locations**
   All 120 V ac receptacle circuits shall either be provided with ground fault circuit interrupter (GFI) protection, or be on a grounded circuit which is periodically tested.

112. **Floor, Floor Openings, Passageways, Stairs**

A. **Floors**
   Floors shall have even surfaces and afford secure footing. Slippery floors or stairs should be provided with antislip covering.

B. **Passageways**
   Passageways, including stairways, shall be unobstructed and shall, where practical, provide at least seven feet headroom.

Where the preceeding requirements are not practical, the obstructions should be painted, marked or indicated by warning signs and the area properly lighted.

C. Railings

All floor openings without gratings or other adequate cover and raised platforms and walkways in excess of one foot in height shall be provided with railings. Openings in railings for units such as fixed ladders, cranes and the like shall be provided with adequate guards such as grates, chains or sliding pipe sections.

D. Stair Guards

All stairways consisting of four or more risers shall be provided with handrails.

NOTE: Additional information may be obtained by referring to American National Standard Safety Code for Floor and Wall Openings, Railings and Toe Boards, A 12.1-1973.

E. Top Rails

All top rails shall be kept unobstructed for a distance of three inches in all directions except from below at supports.

113. Exits

A. Clear Exits

Each room or space and each working space about equipment shall have a means of exit which shall be kept clear of all obstructions. Exit doors shall be equipped with locks or latches that permit opening by means of simple pressure or torque on the actuating parts under any condition.

B. Double Exits

If the plan of the room or space and the character and arrangement of equipment are such that an accident would be likely to close or make inaccessible a single exit, a second exit shall be provided.

114. Fire Extinguishing Equipment

Fire extinguishing equipment approved for the intended use shall be conveniently located and conspicuously marked.

Section 12. Installation and Maintenance of Equipment

120. General Requirements

All electric equipment shall be constructed, installed, and maintained so as to safeguard personnel as far as practical.

121. Inspections

A. In-Service Equipment

Electric equipment shall be periodically inspected and maintained. Defective equipment or wiring shall be put in good order or permanently disconnected.

B. Idle Equipment

Infrequently used equipment or wiring shall be inspected and tested before use to determine its fitness for service. Idle equipment energized but not connected to load shall be inspected or maintained periodically.

C. Emergency Equipment

Equipment or wiring maintained for emergency service shall be periodically inspected and tested to determine its fitness for service.

D. New Equipment

New equipment shall be inspected and tested before being placed in service.

EXCEPTION: The equipment to be tested does not include fittings, devices, appliances, fixtures or other hardware.

122. Guarding Shaft Ends, Pulleys, Belts and Suddenly Moving Parts

A. Mechanical Transmission Machinery

The methods for safeguarding pulleys, belts and other equipment used in the mechanical transmission of power shall be in accordance with American National Standard Safety Standard for Mechanical Power Transmission Apparatus, ANSI B15.1—1972, or the latest revision.

B. Suddenly Moving Parts

Parts of equipment which move suddenly in such a way that persons in the vicinity are likely to be injured by such movement, shall be guarded or isolated.

123. Protective Grounding

A. Protective Grounding or Physical Isolation of Non-Current-Carrying Metal Parts

All electric equipment shall have the exposed noncurrent-carrying metal parts, such as frames of generators and switchboards, cases of transformers, switches and operating levers effectively grounded or physically isolated. All metallic guards including rails, screen fences, etc about electric equipment shall be effectively grounded.

B. Grounding Method

All grounding which is intended to be a permanent and effective protective measure, such as surge arrester grounding, grounding of circuits, equipment, or wire raceways, shall be made in accordance with the methods specified in Section 9 of this code.

NOTE: Additional information is available in IEEE Std 80-1976, Guide for Safety in AC Substation Grounding.

C. Provision for Grounding Equipment During Maintenance

Electric equipment or conductors normally operating at more than 600 V between conductors, on or about which work is occasionally done while isolated from a source of electric energy by disconnecting or isolating switches only, shall be provided with some means for grounding, such as switches, connectors or a readily accessible means for connecting a portable grounding conductor. When necessary, grounding may be omitted on conductors normally operating at 25 kV or less and not influenced by higher voltage conductors, where visible openings in the source of supply are available and are properly tagged in the open position. (See Part 4 of this code.)

124. Guarding Live Parts

A. Where Required

1. Guards shall be provided around all live parts operating above 150 V to ground without an adequate insulating covering, unless their location gives sufficient horizontal or vertical or a combination of these clearances to minimize the possibility of accidental human contact. Clearances from live parts to any permanent supporting surface for workmen shall equal or exceed those shown in Fig 124-1 and shown in Table 124-1.

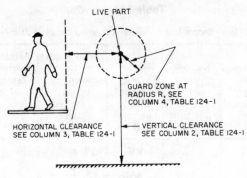

Fig 124-1
Clearance From Live Parts

Table 124-1. Minimum Clearance from Live Parts

PART A — Low, Medium and High Voltages						
Nominal voltage between phases (1)	Minimum vertical clearance of unguarded parts (2)[1]		Minimum horizontal clearance of unguarded parts (3)[1]		Minimum clearance guard to live parts (4)[1]	
	Feet	Inches	Feet	Inches	Feet	Inches
301 to 600	8	8	3	4		2
2,400	8	9	3	4		3
7,200	8	10	3	4		4
13,800	9	0	3	6		6
23,000	9	3	3	9		9
34,500	9	6	4	0	1	0
46,000	9	10	4	4	1	4
69,000	10	5	4	11	1	11
115,000	11	7	6	1	3	1
138,000	12	2	6	8	3	8
161,000	12	10	7	4	4	4
230,000	14	10	9	4	6	4

(Continued on page 94)

Table 124-1   *(Continued)*

PART B — Extra high voltages (based on switching surge factors)②

Maximum design voltage between phases (1)	Switching surge factor③ per unit (A)④	Switching surge line to ground (B)④	Minimum vertical clearance of unguarded parts (2)①		Minimum horizontal clearance of unguarded parts (3)①		Minimum clearance guard to live parts (4)①	
		kV	Ft	In	Ft	In	Ft	In
362,000	2.2 or below	650	15	6	10	0	7	0
	2.3	680	16	0	10	6	7	6
	2.4	709	16	6	11	0	8	0
	2.5	739	17	2	11	8	8	8
	2.6	768	17	9	12	3	9	3
	2.7	798	18	4	12	10	9	10
	2.8	828	18	11	13	5	10	5
	2.9	857	19	7	14	1	11	1
	3.0	887	20	2	14	8	11	8
550,000	1.8 or below	808	18	10	13	4	10	4
	1.9	853	19	6	14	0	11	0
	2.0	898	20	6	15	0	12	0
	2.1	943	21	6	16	0	13	0
	2.2	988	22	6	17	0	14	0
	2.3	1033	23	7	18	1	15	1
	2.4	1078	24	8	19	2	16	2
	2.5	1123	25	10	20	4	17	4
	2.6	1167	27	0	21	6	18	6
	2.7	1212	28	4	22	10	19	10
800,000	1.5	980	22	4	16	10	13	10
	1.6	1045	23	11	18	5	15	5
	1.7	1110	25	6	20	0	17	1
	1.8	1176	27	3	21	9	18	9
	1.9	1241	29	0	23	6	20	6
	2.0	1306	30	10	25	4	22	4
	2.1	1372	32	9	27	3	24	3
	2.2	1437	34	8	29	3	26	2
	2.3	1502	36	9	31	3	28	3
	2.4	1567	38	9	33	3	30	3

(Continued on page 95)

Table 124-1. *(Continued)*

PART C — Extra high voltages (based on BIL factors)[2]

Maximum design voltage between phases (1)	Basic impulse insulation[5] level (BIL) (C)[4]	Minimum vertical clearance of unguarded parts (2)[1]		Minimum horizontal clearance of unguarded parts (3)[1]		Minimum clearance guard to live parts (4)[1]	
	kV	Ft	In	Ft	In	Ft	In
362,000	1050	15	6	10	0	7	0
362,000	1300	17	2	11	8	8	8
550,000	1550	18	10	13	4	10	4
550,000	1800	20	6	15	0	12	0
800,000	2050	22	5	16	11	13	11

Notes and explanations to terms used in Table 124:

(1) Interpolate for Intermediate Values. The clearances in column 4 of this table are solely for guidance in installing guards without definite engineering design and are not to be considered as a requirement for such engineering design. For example, the minimum clearances in the tables above are not intended to refer to the clearances between live parts and the walls of the cells, compartments or similar enclosing structures. They do not apply to the clearances between bus bars and supporting structures nor to clearances between the blade of a disconnecting switch and its base. However, where surge protective devices are applied to protect the live parts, the vertical clearances, Column 2 of Table 124-1 Part A may be reduced provided the clearance is not less than eight feet and six inches plus the electrical clearance between energized parts and ground as limited by the surge protective devices.

(2) Minimum clearances shall satisfy either switching surge or BIL duty requirements, whichever are greater.

(3) Switching Surge Factor — an expression of the maximum Switching Surge Crest Voltage in terms of the maximum operating Line to Neutral Crest Voltage of the power system.

(4) The values of columns A, B, and C are power system design factors that shall correlate with selected minimum clearances. Adequate data to support these design factors should be available.

(5) The selection of station BIL shall be coordinated with surge protective devices when using BIL to determine minimum clearance. BIL—Basic Impulse Insulation Level—See ANSI C92.1-1972 for definition and application.

2. Parts over or near passageways through which material may be carried, or in or near spaces such as corridors, storerooms and boiler rooms used for nonelectrical work shall be guarded or given clearances in excess of those specified such. as may be necessary to secure reasonable safety. The guards shall be substantial and completely shield or enclose the live parts without openings. In spaces used for nonelectrical work, guards should be removable only by means of tools or keys.

3. Parts of indeterminate potential, such as telephone wires exposed to induction from high voltage lines, ungrounded neutral connections, ungrounded frames, ungrounded parts of surge arresters, or ungrounded instrument cases connected directly to a high voltage circuit, shall be guarded on the basis of the maximum voltage which may be present.

B. Strength of Guards

Guards shall be sufficiently strong and shall be supported rigidly and securely enough to prevent them from being displaced or dangerously deflected by a person slipping or falling against them.

C. Types of Guards

1. Location or Physical Isolation

Parts having clearances equal to or greater than specified in Table 124-1 are guarded by location. Parts are guarded by isolation when all entrances to enclosed spaces, runways, fixed ladders, and the like are kept locked, barricaded, or roped off and warning signs are posted at all entrances.

2. Shields or Enclosures

Guards less than four inches outside of the guard zone shall completely enclose the parts from contact up to the heights listed in column 2 of Table 124-1. They shall not be closer to the live parts than listed in column 4 of Table 124-1, except when suitable insulating material is used with circuits of less than 2500 volts to ground. (See note under Table 124-1). If more than four inches outside the guard zone, the guards shall extend a minimum of eight feet, six inches above the floor. Covers or guards, which must at any time be removed while the parts they guard are live, shall be arranged so that they cannot readily be brought into contact with live parts.

3. Railings
   Railings are not substitutes for complete guards. If the
   vertical distance in Table 124-1 cannot be obtained,
   railings may be used. Railings, if used, shall be located at
   a horizontal distance of at least three feet (and preferably not more than four feet) from the nearest point of
   the guard zone which is less than eight feet, six inches
   above the floor. (See Fig 124-2).

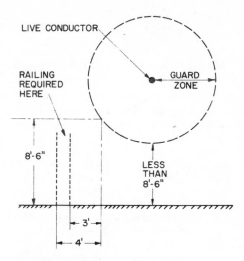

LIVE CONDUCTOR

GUARD ZONE

RAILING REQUIRED HERE

8'-6"

LESS THAN 8'-6"

← 3' →

← 4' →

Fig 124-2
Railing used as Guards

4. Mats
   Mats of rubber or other suitable insulating material
   complying with ANSI/ASTM D178-77, Standard Specification for Rubber Insulating Matting, may be used at
   switchboards, switches, or rotating machinery as supplementary protection.

5. Live Parts Below Supporting Surfaces for Persons
   The supporting surfaces for persons above live parts shall
   be without openings. Toe boards at least six inches high
   and handrails shall be provided at all edges.

6. Insulating Covering on Conductors or Parts
   Conductors and parts may be considered as guarded by insulation if they have either of the following:
   a. Insulation covering of a type and thickness suitable for the voltage and conditions under which they are expected to be operated and if operating above 2500 V to ground having metallic insulation shielding or semiconducting shield in combination with suitable metallic drainage which is grounded to an effective ground.
   b. Barriers or enclosures which are electrically and mechanically suitable for the conditions under which they are expected to be operated.

125. Working Space About Electric Equipment

A. Working Space (600 Volts or Less)
   Access and working space shall be provided and maintained about electric equipment to permit ready and safe operation and maintenance of such equipment.
   1. Clear Spaces
      Working space required by this Section shall not be used for storage. When normally enclosed energized parts are exposed for inspection or servicing, the working space, if in a passageway or general open space, shall be guarded.
   2. Access and Entrance to Working Space
      At least one entrance shall be provided to give access to the working space about electrical equipment.
   3. Working Space
      The working space in the direction of access to energized parts operating at 600 volts or less which require examination, adjustment, servicing, or maintenance while energized shall not be less than indicated in Table 125-1. In addition to the dimensions shown in Table 125-1, the working space shall not be less than 30 inches wide in front of the electric equipment. Distances shall be measured from the energized parts if such are exposed or from the enclosure front or opening if such are enclosed. Concrete, brick, or tile walls shall be considered grounded.
   4. Headroom Working Space
      The minimum headroom of working spaces about switchboards or control centers shall be 7 ft.

Table 125-1   Working Space

Voltage to ground	Minimum clear distance (ft)		
	Condition: 1	2	3
0-150	3	3	3
151-600	3	3½	4

Where the *Conditions* are as follows:

1. Exposed energized parts on one side and no energized or grounded parts on the other side of the working space, or exposed energized parts on both sides effectively guarded by suitable wood or other insulating materials. Insulated wire or insulated busbars operating at not over 300 V shall not be considered energized parts.

2. Exposed energized parts on one side and grounded parts on the other side.

3. Exposed energized parts on both sides of the work space [not guarded as provided in Condition 1] with the operator between.

EXCEPTION: Working space shall not be required in back of assemblies, such as dead-front switchboards or motor control centers where there are no renewable or adjustable parts such as fuses or switches on the back and where all connections are accessible from locations other than the back.

    5.   Front Working Space

      In all cases where there are energized parts normally exposed on the front of switchboards or motor control centers, the working space in front of such equipment shall not be less than 3 ft.

   B.   **Working Space Over 600 Volts**

    Working space shall be in accordance with Table 124-1 clearances for guarding.

126.   Equipment for Work on Energized Parts

When it is necessary for personnel to move themselves, material, or tools within the guard zone of unguarded energized parts, protective equipment shall be provided.

This protective equipment shall be periodically inspected, tested, and kept in a safe condition. Protective equipment shall be rated for not less than the voltage involved. Refer to Part 4 Section A for a list of specifications for equipment.

127. Hazardous Locations

Electrical installations, in hazardous areas shall meet the requirements of articles 500 through 503, and articles 511 through 517 of NFPA-78, the National Electrical Code.

Specific hazardous areas in a power plant or substation and their classifications are identified in the following sections.

A. **Coal-Handling Areas**

1. Tunnels below stockpiles, surge piles, and in, below or above coal storage silos or bunkers or other enclosed coal storage areas where methane may accumulate, are Class I, Division 1, Group D, and Class II, Division 1, Group F locations.

2. Enclosed areas of preparation plants or coalhandling facilities where coal dust might accumulate, are Class II, Division I, Group F locations.

3. Electrical equipment in other locations in which hazardous concentrations of flammable gases or vapors may exist continually, intermittently or periodically under normal operating conditions shall be in accordance with Article 501 of the National Electrical Code or be adequately ventilated.

4. The minimum acceptable requirements for adequate ventilation (pressurization) to reduce the classification of an enclosed area or enclosure within a Class I, Division I area to non-hazardous are:

    a. The ventilation system shall maintain at least 0.1 inch (2.5mm) of positive water pressure in the area with all openings closed.

    b. The ventilation system shall provide a minimum velocity of 60 ft per min outward through each opening with all openings open at the same time.

    c. The ventilation system shall be interlocked so that on failure of the ventilation system, all power to the area shall be deenergized except to those devices which meet the Class I, Division I requirements without the ventilation system.

    d. The maximum operating temperature of any internal surface shall not exceed 80% of the ignition temperature ($^\circ$C) of the hazardous material involved.

5. Locations in which combustible dust is or may be in suspension in the air continuously, intermittently, or periodically under normal operating conditions, or in

quantities sufficient to produce explosions or ignitable mixtures, are classified as Class II, Division 1, Group F locations and all electrical equipment shall be installed and maintained in accordance with the requirements of Article 502 of the National Electrical Code.

6. Locations where dangerous concentrations of suspended dust are prevented during normal operation but where dust accumulations on electrical equipment may be sufficient to interfere with the safe dissipation of heat from electrical equipment or might be ignited by arcs, sparks, or burning material from such equipment are Class II, Division 2, Group F locations and all electrical equipment shall be installed and maintained in accordance with the requirements of Article 502 of the National Electrical Code.

7. Enclosed sections where only wet coal is handled or enclosed sections so cut off as to be free from dangerous amounts of coal dust are non-hazardous. Coal shall be considered to be wet if enough water sprays are installed and maintained to prevent more than 0.3 ounce (.85 g) of coal dust per cu ft (.028 cu m) of enclosed air volume from being thrown into suspension or from accumulating on or in electrical equipment.

8. Locations having completely dust-tight pulverized fuel systems designed and installed in compliance with NFPA 85, shall not be considered hazardous.

9. Portable lamps for use in Fuel Bunkers or Bins shall be suitable for Class II Division 1 locations.

10. Sparking electrical tools shall not be used where flammable dust or dust clouds are present.

11. An equipment grounding conductor shall be carried with the power conductors and serve to ground the frames of all equipment supplied from that circuit. The origin of the grounding conductor shall be:

     a. Ungrounded Delta or Wye — Transformer frame ground.

     b. Grounded Delta or Wye — Transformer grounded secondary connection.

     c. Resistance Grounded Wye — The grounded side of the grounding resistor.

12. Ungrounded systems should be equipped with a ground fault indicating device to give both a visual and audible alarm upon the occurrence of a ground fault in the system.

101

B. **Flammable and Combustible Liquids**

    1. Flammable Liquid shall mean a liquid having a flash point below 100 °F and having a vapor pressure not exceeding 40 pounds per sq in (absolute) at 100 °F and shall be known as a Class I liquid.

    2. Combustible liquid shall mean a liquid having a flash point greater than or equal to 100 °F and having a vapor pressure not exceeding 40 pounds per sq in (absolute) at 100 °F.

    3. Class I liquids are subdivided as follows:

        a. Class IA includes those having flash points below 73 °F and having a boiling point below 100 °F.

        b. Class IB includes those having flash points below 73 °F.

        c. Class IC includes those having flash points at or above 73 °F and below 100 °F.

    4. Combustible liquids are subdivided as follows:

        a. Class II includes those having flash points equal to or greater than 100 °F but less than 140 °F.

        b. Class IIIA includes those having flash points equal to or greater than 140 °F but less than 200 °F.

        c. Class IIIB includes those having flash points greater than or equal to 200 °F.

C. **Flammable Liquid Storage Area**

    1. Electrical wiring and equipment located in inside storage rooms used for Class I liquids shall be approved for Class I, Division 2 locations, (see Table 127-1).

D. **Loading and Unloading Facilities**

Electrical equipment located in the area shall comply with the requirements of Table 127-2.

    1. Static Protection. Bonding facilities for protection against static sparks during the loading of tank vehicles through open domes shall be provided (a) where Class I liquids are loaded, or (b) where Class II or Class III liquids are loaded into vehicles which may contain vapors from previous cargoes of Class I liquids.

        a. Protection as required in 127D1 shall consist of a metallic bond wire permanently electrically connected to the fill stem or to some part of the rack structure in electrical contact with the fill stem. The free end of such wire shall be provided with a clamp or equivalent device for convenient attachment to some metallic part in electrical contact with the cargo tank of the tank vehicle.

Table 127-1 Electrical Equipment Classified Areas —
Flammable Liquid Storage Areas

Location	NEC Class I Division	Extent of Classified Area
Indoor equipment installed where flammable vapor-air mixtures may exist under normal operations.	1	Area within 5 ft of any edge of such equipment, extending in all directions.
	2	Area between 5 ft and 8 ft of any edge of such equipment, extending in all directions. Also, area up to 3 ft above floor or grade level within 5 ft to 25 ft horizontally from any edge of such equipment.*
Outdoor equipment installed where flammable vapor-air mixtures may exist under normal operations.	1	Area within 3 ft of any edge of such equipment extending in all directions.
	2	Area between 3 ft and 8 ft of any edge of such equipment extending in all directions. Also, area up to 3 ft above floor or grade level within 3 ft to 10 ft horizontally from any edge of such equipment.
Tank — Above ground		
Shell, Ends, or Roof and Dike Area	2	Within 10 ft from shell, ends or roof of tank. Area inside dikes to level of top of dike.
Vent	1	Within 5 ft of open end of vent, extending in all directions.
	2	Area between 5 ft and 10 ft from open end of vent, extending in all directions.
Floating Roof	1	Area above the roof and within the shell.

*NOTE: The release of Class I liquids may generate vapors to the extent that the entire building, and possibly a zone surrounding it, should be considered a Class I, Division 2 location.

(Continued on page 104)

Table 127-1   *(Continued)*

Location	NEC Class I Division	Extent of Classified Area
Tank — Underground		
Fill Opening	1	Any pit, box or space below grade level, any part of which is within the Division 1 or 2 classified area.
	2	Up to 18 in above grade level within a horizontal radius of 10 ft from a loose fill connection and within a horizontal radius of 5 ft from a tight fill connection.
Vent — Discharging		
Upward	1	Within 3 ft of open and of vent, extending in all directions.
	2	Area between 3 ft and 5 ft of open end of vent, extending in all directions.
Drum and Container Filling; Outdoors, or Indoors with Adequate Ventilation	1	Within 3 ft of vent and fill opening, extending in all directions.
	2	Area between 3 ft and 5 ft from vent or fill opening, extending in all directions. Also up to 18 in above floor or grade level within a horizontal radius of 10 ft from vent or fill opening.

(Continued on page 105)

Table 127-1   *(Continued)*

Location	NEC Class I Division	Extent of Classified Area
Pumps, Bleeders, Withdrawal Fitting, Meters and Similar Devices		
Indoors	2	Within 5 ft of any edge of such devices, extending in all directions. Also up to 3 ft above floor or grade level within 25 ft horizontally from any edge of such devices.
Outdoors	2	Within 3 ft of any edge of such devices, extending in all directions. Also up to 18 in above grade level within 10 ft horizontally from any edge of such devices.
Pits		
Without Mechanical Ventilation	2	Entire area within pit if any part is within a Division 1 or 2 classified area.
With Mechanical Ventilation	2	Entire area within pit if any part is within a Division 1 or 2 classified area.
Containing Valves, Fittings or Piping, and Not Within a Division 1 or 2 Classified Area	2	Entire Pit
Drainage Ditches, Separators, Impounding Basins	2	Area up to 18 in above ditch, separator or basin. Also up to 18 in above grade within 15 ft horizontally from any edge.

Table 127-2   Electrical Equipment Classified Areas —
Bulk Plants

Location	NEC Class I, Group D Division	Extent of Classified Area
Bottom Loading with Vapor Recovery or Any Bottom Unloading	2	Within 3 ft of point of connections, extending in all directions. Also up to 18 in above grade within a horizontal radius of 10 ft from point of connection.

b. Such bonding connection shall be made fast to the vehicle or tank before dome covers are raised and shall remain in place until filling is completed and all dome covers have been closed and secured.

EXCEPTION: Bonding as specified in 127D1, 127D1a, 127D1b is not required:

(1) where vehicles are loaded exclusively with products not having a static accumulating tendency, such as asphalts including cutback asphalts, most crude oils, residual oils and water soluble liquids;

(2) where no Class I liquids are handled at the loading facility and the tank vehicles loaded are used exclusively for Class II and Class III liquids; and

(3) where vehicles are loaded or unloaded through closed bottom or top connections whether the hose or pipe is conductive or nonconductive.

2. Stray Currents. Tank car loading facilities where flammable and combustible liquids are loaded or unloaded through open domes shall be protected against stray currents by permanently bonding the pipe to at least one rail and to the rack structure, if of metal. Multiple pipes entering the rack area shall be permanently electrically bonded together. In addition, in areas where excessive stray currents are known to exist, all pipe entering the rack area shall be provided with insulating sections to electrically isolate the rack piping from the pipe lines. These precautions are not necessary where Class II or Class III liquids are handled exclusively and there is no probability that tank cars will contain

vapors from previous cargoes of Class I liquids. Temporary bonding is not required between the tank car and the rack or piping during either loading or unloading irrespective of the class of liquid handled.

3. Container Filling Facilities. Class I liquids shall not be dispensed into metal containers unless the nozzle or fill pipe is in electrical contact with the container. This can be accomplished by maintaining metallic contact during filling, by a bond wire between them, or by other conductive path having an electrical resistance not greater than 10^6 ohms. Bonding is not required where a container is filled through a closed system, or is made of glass or other non-conducting material. Recommended Practice on Static Electricity, NFPA No 77, provides information on static protection.

E. Gasoline Dispensing Stations

1. Rule 127E shall apply to areas where Class I liquids are stored, handled or dispensed. For areas where Class II or Class III liquids are stored, handled or dispensed, the electrical equipment may be installed in accordance with the provisions of applicable sections of this code (ANSI C2).

2. All electrical equipment and wiring shall be furnished and installed in accordance with the National Electrical Code NFPA No 70. All electrical equipment integral with the dispensing hose or nozzle shall be suitable for use in Division 1 locations.

3. Table 127-3 shall be used to delineate and classify areas for the purpose of installation of electrical equipment under normal circumstances. A classified area shall not extend beyond an unpierced wall, roof or other solid partition. The designation of classes and divisions is defined in Chapter 5, Article 500, of the National Electrical Code, NFPA No. 70, (ANSI C1).

4. The area classifications listed in Table 127-3 are based on the premise that the installation meets the applicable requirements of this code in all respects. Should this not be the case, the authority having jurisdiction shall have the authority to determine the extent of the classified area.

Table 127-3 Electrical Equipment Classified Areas —
Gasoline Dispensing Stations

Location	NEC Class I Division	Extent of Classified Area
Gasoline Dispensing Units (except overhead type dispensers)	1	The area up to 4 ft vertically above the base within the enclosure or up to a solid partition less than 4 ft above the base, located above the nozzle insertion level and above the level of any gasketed joint, hose or stuffing box.
	2	With 18 in horizontally in all directions from the Division 1 area within the enclosure.
Outdoor	2	Up to 18 in above grade level within 20 ft horizontally of any edge of enclosure.
Indoor with Mechanical Ventilation	2	Up to 18 in above grade for floor level within 20 ft horizontally of any edge of enclosure.
with Gravity Ventilation	2	Up to 18 in above grade or floor level within 25 ft horizontally of any edge of enclosure.
Gasoline Dispensing Units Overhead Type		Within the dispenser enclosure and 18 in in all directions from the enclosure where not suitably cut off by ceiling or wall. All electrical equipment integral with the dispensing hose or nozzle.

(Continued on page 109)

Table 127-3 *(Continued)*

Location	NEC Class I Division	Extent of Classified Area
Gasoline Dispensing Units Overhead Type *(Continued)*	2	An area extending 2 ft horizontally in all directions beyond the Division 1 area and extending to grade below this classified area.
	2	Up to 18 in above grade level with 20 ft horizontally measured from a point vertically below the edge of any dispenser.
Gasoline Dispensing Station Lubrication or Service Room		
With Dispensing	1	Any pit within any unventilated area.
	2	Any pit with ventilation.
	2	Area up to 18 in above floor or grade level and 3 ft horizontally from a lubrication pit.
Dispenser for Class I Liquids	2	With 3 ft of any fill or dispensing point, extending in all directions.
Without Dispensing	2	Entire area within any pit used for lubrication or similar services where Class I liquids may be released.
	2	Area up to 18 in above any such pit, and extending a distance of 3 ft horizontally from any edge of the pit.
Storage and Rest Rooms	Non-classified	If there is any opening to these rooms within the extend of a Division 1 area, the entire room shall be classified as Division 1.

F. Boilers

1. When storing, handling, or burning fuel oils which may have flash points below 100 °F (37.8 °C) (Class I liquids, as defined in NFPA No 30 Flammable and Combustible Liquids Code) or which may be heated above their flash point, attention must be given to electrical installations in areas where flammable vapors or gases may be present in the atmosphere. Typical locations are: burner areas, fuel-handling equipment areas, fuel storage areas, pits, sumps, and low spots where fuel leakage or vapors may accumulate. Article 500 of the National Electrical Code (NFPA No 70) provides for classifying such areas and defines requirements for electrical installations in the areas so classified. The burner front piping and equipment shall be designed and constructed to eliminate hazardous concentrations of flammable gases that exist continuously, intermittently, or periodically under normal operating conditions. Providing the burners are thoroughly purged before removal for cleaning, burner front maintenance operations will not cause hazardous concentrations of flammable vapors to exist frequently. With such provisions, the burner front is not normally classified more restrictively than Class I, Division 2.

2. The operating company shall be responsible for classifying areas where fuel is stored, handled, or burned, and for revising the classification if conditions are changed. Installations shall conform to NFPA No 30, Flammable and Combustible Liquids Code and NFPA No 70 (NEC). Guidance can be obtained from API RP 500, Recommended Practice for Classification of Areas for Electrical Installations in Petroleum Refineries.

G. Gaseous Hydrogen Systems for Supply Equipment

1. Outdoor storage areas shall not be located beneath electric power lines.

2. Safety Considerations at Specific storage areas.
Electrical equipment shall be suitable for Class I, Division 2 locations.

H. **Liquid Hydrogen Systems**
   1. Electrical wiring and equipment located within 3 ft of a point where connections are regularly made and disconnected, shall be in accordance with Article 501 of the National Electrical Code, NFPA 70, for Class I, Group B, Division 1 locations.
   2. Except as provided in Paragraph 11, electrical wiring and equipment located within 25 ft of a point where connections are regularly made and disconnected or within 25 ft of a liquid hydrogen storage container, shall be in accordance with Article 501 of the National Electrical Code, NFPA No 70, for Class I, Group B, Division 2 locations. When equipment approved for Class I, Group B atmospheres is not commercially available, the equipment may be (1) purged or ventilated in accordance with NFPA No 496, Standard for Purged Enclosures for Electrical Equipment in Hazardous Locations, or (2) intrinsically safe, or (3) approved for Class I, Group C atmospheres. This requirement does not apply to electrical equipment which is installed on mobile supply trucks or tank cars from which the storage container is filled.

I. **Sulfur**
   1. Electrical wiring and equipment located in areas where sulfur dust is in suspension in explosive or ignitable mixtures during normal operations, shall be suitable for Class II, Division 1, Group G.

J. **Oxygen**
   Bulk oxygen installations are not defined as hazardous locations.

K. **Liquefied Petroleum Gas (LPG)**
   1. LPG is heavier than air.
   2. Since LPG is contained in a closed system of piping and equipment, the system need not be electrically conductive or electrically bonded for protection against static electricity.
   3. Fixed electrical equipment and wiring installed within classified areas specified in Table 127-4 shall meet the requirements of Chapter 5, Article 500 of NFPA No 70 (ANSI C1).

**Table 127-4   Electrical Equipment Classified Areas —
LPG Storage**

Location	NEC Class I Group D	Extent of Classified Area
Storage Containers other than DOT Cylinders	2	Within 15 ft in all directions from connections, except connections otherwise covered in Table K-1.
Tank Vehicle and Tank Car Loading and Unloading	1	Within 5 ft in all directions from connections regularly made or disconnected for product transfer.
	2	Beyond 5 ft but within 15 ft in all directions from a point where connections are regularly made or disconnected and within the cylindrical volume between the horizontal equator of the sphere and grade.
Gage Vent Openings other than those on DOT Cylinders	1	Within 5 ft in all direction from point of discharge.
	2	Beyond 5 ft but within 15 ft in all directions from point of discharge.
Relief Valve Discharge other than those on DOT Cylinders	1	Within direct path of discharge. Note: Fixed electrical equipment should preferably not be installed.
	1	Within 5 ft in all directions from point of discharge.
	2	Beyond 5 ft but within 15 ft in all directions from point of discharge except within the path of discharge.

(Continued on page 113)

Table 127-4 *(Continued)*

Location	NEC Class I Group D	Extent of Classified Area
Pits or trenches containing or located beneath LP-Gas valves, regulators, and similar equipment		
Without mechanical ventilation	1	Entire pit or trench.
	2	Entire room and any adjacent room not separated by a gastight partition.
	2	With 15 ft in all directions from pit or trench when located outdoors.
With adequate mechanical ventilation	2	Entire pit or trench.
	2	Entire room and any adjacent room not separated by a gastight partition.
	2	With 15 ft in all directions from pit or trench when located outdoors.
Special Buildings or rooms for storage of portable containers	2	Entire room.
Pipelines and connections containing operational bleeds, drips, vents or drains	1	Within 5 ft in all directions from point of discharge.
Container Filling:		
Indoors without ventilation	1	Entire room.
Indoor with adequate ventilation	1	Within 5 ft in all directions and connections regularly made or disconnected for product transfer.
	2	Beyond 5 ft and entire room.

(Continued on page 114)

Table 127-4 *(Continued)*

Location	NEC Class I Group D	Extent of Classified Area
Container Filling:		
Outdoors in open air	1	Within 5 ft in all directions and connections regularly made or disconnected for product transfer.
	2	Beyond 5 ft but within 15 ft in all directions from a point where connections are regularly made or disconnected and within the cylindrical volume between the horizontal equator or the sphere and grade.

L. Natural Gas (Methane)
1. Natural Gas is lighter than air.
2. Since Natural Gas is contained in a closed system of piping and equipment, the system need not be electrically conductive or electrically bonded for protection against static electricity.
3. Fixed electrical equipment and wiring installed within classified areas specified in Table 127-5 shall meet the requirements of Chapter 5, Article 500 of the NFPA No 70 (ANSI C1).

128. Identification

Electrical equipment and devices shall be identified for safe use and operation. The identification shall be as nearly uniform as practical throughout any one station. Identification marks shall not be placed on removable covers or doors where the interchanging of those covers or doors is possible.

Table 127-5   Electrical Equipment Classified Areas —
Natural Gas (Methane) Areas

Location	NEC Class I Group D	Extent of Classified Area
Nonfired areas containing gas pipeline connections, valves or gages		
Indoors with adequate ventilation	2	Entire room and any adjacent room not separated by a gastight partition and 15 ft beyond any wall or roof ventilation discharge vent or louver.
Outdoors in open air at or above grade	2	Within 15 ft in all directions of connections valves or gages.
Pits, Trenches or Sumps located in or adjacent to Division 1 or 2 areas	1	Entire pit, trench or sump.

Section 13. Rotating Equipment

Rotating equipment includes generators, motors, motor generators and rotary converters.

130. Speed Control and Stopping Devices

A. Automatic Overspeed Trip Device for Prime Movers
When harmful overspeed can occur, prime movers driving generating equipment shall be provided with automatic overspeed trip devices in addition to their governors.

B. Manual Stopping Devices
Stopping devices, such as switches or valves which can be operated from locations convenient to machine operators, shall be provided for all prime movers and for motors driving generating equipment.

Manual controls to be used in emergency for machinery and electrical equipment shall be located so as to provide protection to the operator during such emergency.

C. Speed Limit for Motors
Machines of the following types shall be provided with speed-limiting devices unless their inherent characteristics or the load and the mechanical connection thereto are such as to safely limit the speed.
1. Separately excited direct-current motors.
2. Series motors.

D. Low-Voltage Protection of Motors
All motors so employed or arranged that an unexpected starting of the motor is a personnel hazard shall be equipped with low-voltage protection. This shall automatically cause and maintain the interruption of the motor circuit when the voltage falls below an operating value. This rule does not apply to those motors with an emergency use and where the opening of the circuit may cause less safe conditions.

E. Adjustable-Speed Motors
Adjustable-speed motors, controlled by means of field regulation, shall, in addition to the provisions of Rule 130C, be so equipped and connected that the field cannot be weakened sufficiently to permit dangerous speed.

F. Protection of Control Circuits

Where speed-limiting or stopping devices and systems are electrically operated, the control circuits by which such devices are actuated shall be protected from mechanical damage. Such devices and systems should be of the automatic tripping type.

131. Motor Control

If the starting is automatic, as for example, by a float switch, or if the starting device or control switch is not in sight, or more than 50 ft distant from the motor and all parts of the machinery operated, the power or control circuit shall be such that it can positively be kept open such as by use of lockout/tagout procedures.

132. Mobile Hydrogen Equipment

Mobile hydrogen supply units being used to replenish a hydrogen system shall be bonded both to the grounding system and to the grounded parts of the hydrogen system.

Section 14. Storage Batteries

140. General

The provisions of this section are intended to apply to all stationary installations of storage batteries. For operating precautions, see Part 4 of this code.

Space shall be provided around batteries for safe inspection, maintenance, testing, and cell replacement and space left above the cells to allow for operation of lifting equipment when required, addition of water, and taking measurements.

141. Location

Storage batteries shall be located within a protective enclosure or area accessible only to qualified persons. A protective enclosure can be a battery room, control building, or a case, cage, or fence which will protect the contained equipment and minimize the possibility of inadverdent contact with energized parts.

142. Ventilation

The battery area shall be ventilated, either by a natural or powered ventilation system to prevent accumulation of hydrogen. The ventilation system shall limit hydrogen accumulation to less than an explosive mixture. Failure of continuously operated or automatically controlled powered ventilation system shall be annunciated.

143. Racks

Racks refer to frames designed to support cells or trays. Racks shall be firmly anchored preferably to the floor. Anchoring to both walls and floors is not recommended.

Racks made of metal shall be grounded.

144. Floors in Battery Areas

Floors of battery areas should be an acid-resistive material, or be painted with acid-resistive paint, or otherwise protected. Provision should be made to contain spilled electrolyte.

145. Illumination for Battery Areas

Lighting fixtures shall be protected from physical damage by guards or isolation. Receptacles and lighting switches should be located outside of battery areas.

146. Service Facilities

A. Proper eye protection and clothing shall be provided in the battery area during battery maintenance and installation and shall consist of:
1. Goggles or face shield
2. Acid resistant gloves
3. Protective aprons and overshoes
4. Portable or stationary water facilities for rinsing eyes and skin.
5. Neutralizing agent.

B. Warning signs inside and outside of a battery room or in the vicinity of a battery area, prohibiting smoking, sparks or flame shall be provided.

147. Battery Charger

It is recommended that the battery charger and associated control equipment be located outside a battery room or away from the vicinity of a battery area.

Section 15. Transformers and Regulators

150. Current-Transformer Secondary Circuits Protection When Exceeding 600 Volts

Secondary circuits, when in a primary voltage area exceeding 600 V should, except for short lead lengths at the terminals of the transformer, have the secondary wiring adequately protected by means of grounded conduit or by a grounded metallic covering. Current transformers shall have provision for shorting the secondary winding.

151. Grounding Secondary Circuits of Instrument Transformers

The secondary circuits of instrument transformers shall be effectively grounded where functional requirements permit.

152. Location and Arrangement of Power Transformers and Regulators

A. Outdoor Installations
   1. A transformer or regulator shall be installed so that all energized parts are enclosed or guarded so as to minimize the possibility of inadverdent contact, or the energized parts shall be isolated in accordance with Rule 124. The case shall be grounded in accordance with Rule 123.
   2. Oil-filled transformers shall be protected by one or more of the following methods to minimize fire hazards. The method to be applied shall be according to the degree of fire hazard and the amount of oil contained in the transformer. Recognized methods are space separation, fire-resistant barriers, automatic extinguishing systems, absorption beds and enclosures.

      The amount of oil contained should be considered in the selection of space separation, fire resistant barriers, automatic extinguishing systems, absorption beds, and enclosures which confine the oil of a ruptured transformer tank all of which are recognized safeguards.

B. Indoor Installations
   1. Transformers and regulators 75 kVA and above containing an appreciable amount of flammable liquid and

located indoors shall be installed in ventilated rooms or vaults separated from the balance of the building by fire walls. Doorways to the interior of the building shall be equipped with fire doors and shall have means of containing the oil.

2. Transformers or regulators of the dry type or containing a non flammable liquid or gas may be installed in a building without a fireproof enclosure. When installed in a building which is used for other than station purposes the case or the enclosure shall be designed so that all energized parts are enclosed in the case grounded in accordance with Rule 123. As an alternate, the entire unit may be enclosed so as to minimize the possibility of inadverdent contact by persons with any part of the case or wiring. When installed, the pressure relief vent of a unit containing a non-biodegradable liquid shall be furnished with a means for absorbing toxic gases.

Section 16. Conductors

160. Electrical Protection

Conductors shall be suitable for the location, use and voltage.

A. Overcurrent Protection Required

Conductors and insulation shall be protected against excessive heating by the design of the system and by overcurrent, alarm, indication, or trip devices.

B. Grounded Conductors

Conductors normally grounded for the protection of persons shall be arranged without overcurrent protection or other means which could interrupt their continuity to ground.

161. Mechanical Protection

All conductors shall be adequately supported to withstand forces caused by the maximum short circuit current to which they may be subjected.

Where exposed to mechanical damage, casing, armor, or other means shall be employed to prevent damage or disturbance to conductors, their insulation, or supports.

162. Isolation

All non-shielded insulated conductors of more than 2500 volts to ground and bare conductors of more than 150 V to ground, shall be isolated by elevation or guarded in accordance with Rule 124.

Non-shielded, insulated, and jacketed conductors may be installed in accordance with Rule 124C6.

163. Conductor Terminations

A. Insulation

Ends and joints of insulated conductors, unless otherwise adequately guarded, shall have insulating covering equivalent to that of other portions of the conductor.

B. Metal-Sheathed or Shielded Cable

Insulation of the conductors where leaving the metal sheath or shield, shall be protected from mechanical damage, moisture and excessive electrical stress.

Section 17. Circuit Breakers, Reclosers, Switches and Fuses

170. Arrangement

Circuit breakers, reclosers, switches and fuses shall be so installed as to be accessible only to persons qualified for operation and maintenance. Walls, barriers, latched doors, location, isolation or other means shall be provided to protect persons from energized parts or arcing. Conspicuous marking shall be provided at the device and at any remote operating points to identify the equipment controlled. When the contact parts of a switching device are not normally visible, the device shall be equipped with an indicator to show all normal operating positions.

171. Application

Circuit breakers, reclosers, switches, and fuses should be utilized with due regard to their assigned ratings of voltage and continuous and momentary currents. Circuit breakers, reclosers and fuses which perform a fault current interrupting function shall be capable of safely interrupting the maximum short circuit current available from the system at the point of application. The interrupting capacity should be reviewed prior to each significant system change.

172. Circuit Breakers, Reclosers and Switches Containing Oil

Circuit interrupting devices containing flammable liquids shall be adequately segregated from other equipment and buildings to limit damage in the event of an explosion or fire. Segregation may be provided by spacing, by fire resistant barrier walls, or by metal cubicles. Gas relief vents should be equipped with oil separating devices or piped to a safe location. Means shall be provided to control oil which could be discharged from vents or by tank rupture. This may be accomplished by absorption beds, pits, drains, or by any combination of these. Buildings or rooms housing this equipment shall be of fire resistant construction.

173. Switches and Disconnecting Devices

A. Capacity

Switches shall be of suitable voltage and ampere rating for the circuit in which they are installed. Switches used to break load current shall be marked with the current which they are rated to interrupt. It is recommended that switches that are not rated to interrupt the full load of the circuit be interlocked with circuit breakers to minimize the possibility of the switches being opened under load.

B. Provisions for Disconnecting

Switches and disconnectors shall be so arranged that they can be locked in the open and closed positions, or plainly tagged where it is not possible to install locks. For devices that are operated remotely and automatically, the control circuit shall be provided with a positive disconnecting means near the apparatus to prevent accidental operation of the mechanism.

C. Visible Break Switch

A visible break switch or disconnector shall be inserted in each ungrounded conductor between electric supply equipment or lines and sources of energy of more than 600 V, if the equipment or lines may have to be worked on without protective grounding while the sources may be energized.

Where metal clad switchgear equipment is used, the withdrawn position of the circuit breaker, where clearly indicated, constitutes a visible break for this purpose.

174. Disconnection of Fuses

Fuses in circuits of more than 150 V to ground or more than 60 A shall be classified as disconnecting fuses or be arranged so that before handling:

A. The fuses can be disconnected from all sources of electric energy, or
B. The fuses can be conveniently removed by means of insulating handles.

Fuses can be used to disconnect from the source when they are so rated.

Section 18. Switchgear and Metal Enclosed Bus

180. Switchgear Assemblies

A. General Requirements for All Switchgear

1. Switchgear shall be anchored to minimize movement.
2. Cable routed to switchgear shall be supported in a manner that prevents tension being applied to the conductor terminals.
3. Piping containing liquids, or corrosive or hazardous gases, shall not be routed in the vicinity of switchgear unless suitable barriers are installed to protect the switchgear from damage in the event of a pipe failure.
4. Switchgear shall not be located where flammable or corrosive gases or liquids could enter the enclosure.
5. Switchgear should not be installed in a location which is still specifically under active construction, especially where welding and burning are required directly overhead. Special precautions should be observed to minimize impingement of slag, metal filings, moisture, dust, or hot particles.

   EXCEPTION: Switchgear may be installed in a general construction area provided suitable temporary protection is provided to minimize the risks associated with general construction activities.

6. Steps shall be taken to assure that switchgear is deenergized prior to performing maintenance involving removal of the protective barriers.
7. Steps shall be taken to assure that all personnel safety features are replaced after maintenance work is completed prior to returning the equipment to normal operation.

8. Precautions shall be taken to protect energized switch-gear from damage when maintenance is performed in the area.

9. Switchgear enclosure surfaces shall not be used as physical support for any item unless specifically designed for that purpose.

10. Enclosure interiors shall not be used as storage areas unless specifically designed for the purpose.

11. Metal instrument cases shall be grounded, enclosed in covers which are metal and grounded, or of insulating material.

B. Metal Enclosed Power Switchgear

1. Switchgear shall not be located within 25 ft horizontally indoors or 10 ft outdoors of storage containers, vessels, utilization equipment or devices containing flammable liquids or gases.

EXCEPTION: If an intervening barrier, designed to mitigate the potential effects of flammable liquids or gases, is installed, the distances listed above do not apply.

2. Enclosed switchgear rooms shall have at least two means of egress, one at each extreme of the area, not necessarily in opposite walls. Doors shall swing out and be equipped with locks or latches that permit opening by means of simple pressure or torque on the actuating parts under any condition.

3. Space shall be maintained in front of switchgear to allow breakers to be removed and turned without obstruction.

4. Space shall be maintained in the rear of the switchgear to allow for door opening to at least 90° open, or a minimum of 3 feet and no inches without obstruction when removable panels are used.

5. Permanently mounted devices, panelboards, etc located on the walls shall not encroach on the space requirements in 181B4.

6. Where columns extend into the room beyond the wall surface, the face of the column shall not encroach on the space requirements in 181B4.

7. Low-voltage cables or conductors, except those to be connected to equipment within the compartment, shall not be routed through the medium- or high-voltage divisions of switchgear.

8. Low voltage conductors routed from medium or high voltage sections of switchgear shall terminate in a low voltage section before being routed external to the switchgear.

9. Conductors entering switchgear shall be insulated for the higher operating voltage in that compartment or be separated from insulated conductors of other voltage ratings.

10. Switchgear enclosures shall be suitable for the environment in which they are installed.

11. A warning sign shall be placed in each cubicle containing more than one high voltage source.

12. The location of control devices shall be readily accessible to personnel. Instruments, relays and other devices requiring reading or adjustments should be so placed that work can readily be performed from the working space.

13. Switchgear shall be made of noncombustible and moisture resistant material.

C. Dead Front Power Switchboards

Dead front power switchboards with uninsulated rear connections shall be installed in rooms or spaces that are capable of being locked, with access limited to qualified personnel.

D. Motor Control Centers

1. Motor control centers shall not be connected to systems having higher short circuit capability than the bus bracing can withstand. Where current limiting fuses are employed on the source side of the bus, the bus bracing and breaker interrupting rating are determined by the peak let-through characteristic of the current limiting fuse.

2. A warning sign shall be placed in each cubicle containing more than one voltage source.

E. Control Switchboards

1. Cabinets containing solid-state logic devices, electron tubes or relay logic devices such as boiler analog, burner safety, annunciators, computers, inverters, precipitator logic, soot blower control, load control, telemetering, totalizing microwave radio, etc are covered under these rules.

2. Where carpeting is installed in rooms containing control switchboards, it shall be antistatic type and shall minimize the release of noxious, corrosive, caustic, or toxic gas under any condition.

3. Layout of the installation shall provide adequate clearance in front of, or rear of panels if applicable, to allow meters to be read without use of stools or auxiliary devices.

4. Where personnel access to control panels such as benchboards is required, cables shall be routed through openings separate from the personnel opening. Removable, sliding, or hinged panels are to be installed to close the personnel opening when not in use.

181. Metal Enclosed Bus

A. General Requirements for All Types of Bus

1. Busways shall be installed only in accessible areas.

2. Busways unless specifically approved for the purpose, shall not be installed: where subject to severe physical damage or corrosive vapors; in hoistways; in any classified hazardous location; outdoors or in damp locations.

3. Dead ends of busway shall be closed.

4. Busways should be marked with the voltage and current rating for which they are designed, in such manner as to be visible after installation.

B. Isolated Phase Bus

1. The minimum clearance between an isolated-phase bus and any magnetic material shall be the distance recommended by the manufacturer to avoid overheating of the magnetic material.

2. Non-magnetic conduit should be used to protect the conductors for bus alarm devices, thermocouples, space heaters, etc if routed within the manufacturer's recommended minimum distance to magnetic material and parallel to isolated-phase bus enclosures.

3. When enclosure drains are provided for isolated-phase bus, necessary piping shall be provided to divert water away from electrical equipment.

4. Wall plates for isolated-phase bus shall be non-magnetic, such as aluminum or stainless steel.

5. The generator isolated-phase bus grounding conductor shall be grounded at one point only. Enclosures of bus

and accessories such as potential transformers, surge arresters, capacitors, and generator neutral grounding equipment shall be grounded with a conductor adequately sized to carry the bus momentary current rating for two seconds.

6. Grounding conductors for isolated-phase bus accessories should not be routed through ferrous conduit.

Section 19. Surge Arresters

190. General Requirements

If arresters are required, they shall be located as close as practical to the equipment they protect.

191. Indoor Locations

Arresters, if installed inside of buildings shall be enclosed or shall be located well away from passageways and combustible parts.

192. Grounding Conductors

Grounding conductors shall be run as directly as possible between the arresters and ground and be of low impedance and ample current-carrying capacity (see Section 9 for methods of protective grounding).

193. Installation

Arresters shall be installed in such a manner and location that neither the expulsion of gases nor the arrester disconnector is directed upon live parts in the vicinity.

American National Standard

National Electrical Safety Code

Part 2 (Sections 20-28). Safety Rules for the Installation and Maintenance of Overhead Supply and Communication Lines

2

129

PART 2. Safety Rules for the Installation and Maintenance of Overhead Electric Supply and Communication Lines

Section 20. Purpose, Scope, and Application of Rules

200. Purpose

The purpose of Part 2 of this code is the practical safeguarding of persons during the installation, operation, or maintenance of overhead supply and communication lines and their associated equipment.

201. Scope

Part 2 of this code covers supply and communication conductors and equipment in overhead lines. It covers the associated structural arrangements of such systems and the extension of such systems into buildings. The rules include requirements for spacing, clearances, and strength of construction. They do not cover installations in electric supply stations.

202. Application of Rules

The general requirements for application of these rules are contained by Rule 013. However, when a structure is replaced, the arrangement of equipment shall conform to the current edition of Rule 238C.

Section 21. General Requirements

210. Rule 210 not used in this edition.

211. Rule 211 not used in this edition.

212. Induced Voltages

Rules covering supply line influence and communication line susceptiveness have not been detailed in this code. Cooperative procedures are recommended in the control of voltages induced from proximate facilities. Therefore, reasonable advance notice should be given to owners or operators of other proximate facilities which may be adversely affected by new construction or changes in existing facilities.

213. Accessibility

All parts which must be examined or adjusted during operation shall be arranged so as to be accessible to authorized persons by the provision of adequate climbing spaces, working spaces, working facilities, and clearances between conductors.

214. Inspection and Tests of Lines and Equipment

A. When In Service

1. Initial Compliance With Rules

Lines and equipment shall comply with these safety rules when placed in service.

2. Inspection

Lines and equipment shall be inspected from time to time at such intervals as experience has shown to be necessary.

3. Tests

When considered necessary, lines and equipment shall be subjected to practical tests to determine required maintenance.

4. Record of Defects

Any defects affecting compliance with this code revealed by inspection or tests, if not promptly corrected, shall be recorded; such records shall be maintained until the defects are corrected.

5. Remedying Defects

Lines and equipment with recorded defects which would endanger life or property shall be promptly repaired, disconnected, or isolated.

B. When Out of Service

1. Lines infrequently used

Lines and equipment infrequently used shall be inspected or tested as necessary before being placed into service.

2. Lines temporarily out of service

Lines and equipment temporarily out of service shall be maintained in a safe condition.

3. Lines permanently abandoned

Lines and equipment permanently abandoned shall be removed or maintained in a safe condition.

215. Grounding of Circuits, Supporting Structures, and Equipment

A. Methods

Grounding required by these rules shall be in accordance with the applicable methods given in Section 9.

B. Circuits

1. Common Neutral

A conductor used as a common neutral for primary and secondary circuits shall be effectively grounded as specified in Section 9.

2. Other Neutrals

Primary or secondary neutral conductors, other than common neutrals, which are to be effectively grounded, shall be grounded as specified in Section 9.

3. Surge Arresters

Where the operation of surge arresters is dependent upon grounding, they shall be grounded in accordance with the methods outlined in Section 9.

4. Use of Earth as Part of Circuit

Supply circuits shall not be designed to use the earth normally as the sole conductor for any part of the circuit.

C. Noncurrent-Carrying Parts

1. General

Metal or metal reinforced supporting structures, including lamp posts; metal conduits and raceways; cable sheaths; messengers; metal frames, cases and hangers of equipment; and metal switch handles and operating rods shall be effectively grounded.

EXCEPTION 1: This rule does not apply to frames, cases, and hangers of equipment and switch handles and operating rods which are 8 feet or more above readily accessible surfaces or are otherwise isolated or guarded and where the practice of not grounding such items has been a uniform practice over a well defined area.

EXCEPTION 2: This rule does not apply to isolated or guarded equipment cases in certain specialized applications, such as series capacitors where it is necessary that equipment cases be either ungrounded or connected to the circuit. Such equipment cases shall be considered as energized and shall be suitably identified.

EXCEPTION 3: This rule does not apply to equipment cases, frames, equipment hangers, conduits, raceways, and cable sheaths enclosing only communications conductors, provided they are not exposed to probable contact with open supply conductors of over 300 volts.

2. Guys

Guys shall be effectively grounded if attached to a supporting structure carrying any supply conductor of more than 300 volts or if exposed to such conductors.

EXCEPTION 1: This rule does not apply to guys containing an insulator or insulators installed in accordance with and meeting the requirements of Rule 283.

EXCEPTION 2: This rule does not apply to guys attached to supporting structures if all supply conductors are in cable conforming to the requirements of Rules 230C1, 230C2, or 230C3.

EXCEPTION 3: This rules does not apply if the guy is attached to a supporting structure on private right-of-way if all the supply circuits exceeding 300 volts meet the requirements of Rule 220B2.

216. Arrangement of Switches

A. Accessibility

Switches or their control mechanisms shall be installed so as to be accessible to authorized persons.

B. Indicating Open or Closed Position

Switch position shall be visible or clearly indicated.

C. Locking

Switch operating mechanisms which are accessible to unauthorized persons shall have provisions for locking in each operational position.

D. Uniform Position

The handles or control mechanisms for all switches throughout any system should have consistent positions when opened and uniformly different positions when closed in order to minimize operating errors. Where this practice is not followed, the switches should be marked to minimize mistakes in operation.

Section 22. Relations Between Various Classes of Lines

220. Relative Levels

A. Standardization of Levels

The levels at which different classes of conductors are to be located should be standardized by agreement of the utilities concerned.

B. Relative Levels: Supply and Communication Conductors

  1. Preferred Levels

Where supply and communication conductors cross each other or are located on the same structures, the supply conductors should be carried at the higher level.

EXCEPTION: This rule does not apply to trolley feeders which may be located for convenience approximately at the level of the trolley-contact conductor.

  2. Special Construction for Supply Circuits, the Voltage of Which is 600 Volts or Less and Carrying Power Not in Excess of 5 Kilowatts

Where all circuits are owned or operated by one party or where cooperative consideration determines that tne circumstances warrant and the necessary coordinating methods are employed, single-phase alternating-current or two-wire direct-current circuits carrying a voltage of 600 volts or less between conductors, with transmitted power not in excess of 5 kilowatts, when involved in the joint use of structures with communication circuits may be installed in accordance with footnote 14 of Table 232-1 and footnote 1 of Table 235-5, under the following conditions.

    a. That such supply circuits are of covered conductor not smaller than No. 8 AWG medium hard-drawn copper or its equivalent in strength, and the construction otherwise conforms with the requirements for supply circuits of the same class.

    b. That the supply circuits be placed on the end and adjacent pins of the lowest through signal support arm and that a 30 inch climbing space be maintained from the ground up to a point at least 24 inches above the supply circuits. The supply circuits shall be rendered conspicuous by the use of insulators of different form or color from others on the pole line or by stenciling the voltage on each side of the support arm between the pins carrying each supply circuit, or by indicating the voltage by means of metal characters.

    c. That there shall be a vertical clearance of at least 2 feet between the support arm carrying these supply circuits and the next support arm above. The other pins on the support arm carrying the supply circuits may be occupied by communication circuits used in the operation or control of signal system or other supply system if owned, operated, and maintained by the same company operating the supply circuits.

d. That such supply circuits shall be equipped with arresters and fuses installed in the supply end of the circuit and where the signal circuit is alternating current, the protection shall be installed on the secondary side of the supply transformer. The arresters shall be designed so as to break down at approximately twice the voltage between the wires of the circuit, but the breakdown voltage of the arrester need not be less than 1 kilovolt. The fuses shall have a rating not in excess of approximately twice the maximum operating current of the circuit, but their rating need not be less than 10 amperes. The fuses likewise shall in all cases have rating of at least 600 volts, and where the supply transformer is a stepdown transformer, shall be capable of opening the circuit successfully in the event the transformer primary voltage is impressed upon them.

e. Such supply circuits in cable meeting the requirements of Rules 230C1, 230C2, or 230C3 may be installed below communication attachments, with not less than 2 feet vertical separation between the supply cable and the lowest communication attachment. Communication circuits other than those used in connection with the operation of the supply circuits shall not be carried in the same cable with such supply circuits.

f. Where such supply conductors are carried below communication conductors, transformers and other apparatus associated therewith shall be attached only to the sides of the support arm in the space between, and at no higher level than, such supply wires.

g. Lateral runs of such supply circuits carried in a position below the communication space shall be protected through the climbing space by wood molding or equivalent covering, or shall be carried in insulated multiple-conductor cable, and such lateral runs shall be placed on the under side of the support arm.

C. Relative Levels: Supply Lines of Different Voltage Classifications (as classified in Table 235-5)

1. At Crossings or Conflicts
Where supply conductors of different voltage classifica-

tions cross each other or structure conflict exists, the higher voltage lines should be carried at the higher level.

2. On Structures Used Only by Supply Conductors

Where supply conductors of different voltage classifications are on the same structures, relative levels should be as follows:

a. Where all circuits are owned by one utility, the conductors of higher voltages should be placed above those of lower voltage.

b. Where different circuits are owned by separate utilities, the circuits of each utility may be grouped together and one group of circuits may be placed above the other group provided that the circuits in each group are located so that those of higher voltage are at the higher levels and that any of the following conditions is met:

(1) A vertical spacing of not less than that required by Table 235-5 is maintained between the nearest line conductors of the respective utilities.

(2) Conductors of a lower voltage classification placed at a higher level than those of a higher classification shall be placed on the opposite side of the structure.

(3) Ownership and voltage are prominently displayed.

221. Avoidance of Conflict

Two separate lines, either of which carries supply conductors, should be so separated from each other that neither conflicts with the other. If this is not practical, the conflicting line or lines should be separated as far as possible and shall be built to the grade of construction required by Section 24 for a conflicting line, or the two lines shall be combined on the same structures.

222. Joint Use of Structures

Joint use of structures should be considered for circuits along the same general route. The choice between joint use of structures and separate lines shall be determined through cooperative consideration of all the factors involved, including the character of circuits, the total number and weight of conductors, tree conditions, number and location of branches and service drops, possible structure conflicts, availability of right-of-way, etc. Where such joint use is mutually agreed upon, it shall be subject to the appropriate grade of construction specified in Section 24.

Section 23. Clearances

230. General

A. Application

This section covers all clearances, including climbing spaces, involving overhead supply and communications lines. Clearances of equipment from structure surfaces, from spaces accessible to the general public, and height above ground are covered in Rule 286.

B. Measurement of Clearance and Spacing

Unless otherwise stated, all clearances shall be measured from surface to surface and all spacings shall be measured center to center. For clearance measurement, live metallic hardware electrically connected to line conductors shall be considered a part of the line conductors. Metallic bases of potheads, surge arresters, and similar devices shall be considered a part of the supporting structure.

C. Supply Cables

For clearance purposes, supply cables, including splices and taps, conforming to any of the following requirements are permitted lesser clearances than bare conductors of the same voltage. Cables should be capable of withstanding tests applied in accordance with an applicable standard.

1. Cables of any voltage having an effectively grounded continuous metal sheath or shield, or cables designed to operate on a multi-grounded system at 8.7 kV or less, having a semiconducting insulation shield in combination with suitable metallic drainage, all supported on and cabled together with an effectively grounded bare messenger-neutral.

2. Cables of any voltage, not included in 230C1, covered with a continuous auxiliary semiconducting shield in combination with suitable metallic drainage and supported on and cabled together with an effectively grounded bare messenger.

3. Insulated, non-shielded cable operated at not over 5 kV phase-to-phase, or 2.9 kV phase-to-ground, supported on and cabled together with an effectively grounded bare messenger.

D. Covered Conductors

Covered conductors shall be considered bare conductors for all clearance requirements except that spacing between conductors of the same or different circuits, including grounded conductors, may be reduced below the minimum requirements for bare conductors when the conductors are owned, operated, or maintained by the same party and when the conductor covering provides sufficient dielectric strength to prevent a short circuit in case of momentary contact between conductors or between conductors and the grounded conductor. Intermediate spacers may be used to maintain conductor spacing and provide support.

E. Neutral Conductors

1. Neutral conductors which are effectively grounded throughout their length and associated with circuits of 0 to 22 kilovolts to ground may have the same clearances as guys and messengers, except as provided for conductors over railroads in Rule 232A, Table 232-1, footnote 15.

2. All other neutral conductors of supply circuits shall have the same clearances as the phase conductors of the circuit with which they are associated.

F. Alternating and Direct Current Circuits

The rules of this section are applicable to both alternating and direct current circuits. For direct current circuits, the clearance requirements shall be the same as those for alternating current circuits having the same crest voltage to ground.

G. Constant-Current Circuits

The clearances for constant-current circuits shall be determined on the basis of their nominal full-load voltage.

H. Maintenance of Clearances and Spacings

The clearances and spacing required shall be maintained at the values and under the conditions specified in Section 23.

231. Clearances of Supporting Structures from Other Objects

Supporting structures, support arms and equipment attached thereto, and braces shall have the following clearances from other objects. The clearance shall be measured between the nearest parts of the objects concerned.

A. **From Fire Hydrants**
Not less than 3 ft.

RECOMMENDATION: Where conditions permit, a clearance of not less than 4 ft is recommended.

B. **From Streets, Roads, and Highways**
1. Where there are curbs, the measurement from the street side of the curb to the supporting structure, support arms, and equipment attached thereto up to 15 ft above the road surface.
    a. *shall be* not less than 6 in.
    b. *should be* not less than 2 ft on arterials which are primarily used for through traffic, usually on a continuous route.
    c. *should be* not less than 1 ft on local streets and roads which are primarily used for access to residences, businesses, or other abutting properties.

    RECOMMENDATION: Where sufficient border space is available within the street, road, or highway right-of-way, placement of supporting structures as far as is practical behind the curbs is recommended. Location behind sidewalks is desirable.

2. Where there are no curbs, supporting structures should be at or as near as is practical to the street, road, or highway right-of-way line.
3. Location of overhead utility installations on highways with narrow rights-of-way or on urban streets with closely abutting improvements are special cases which must be resolved in a manner consistent with the prevailing limitations and conditions.

C. **From Railroad Tracks**
Where railroad tracks are paralleled or crossed by overhead lines, all portions of the supporting structures, support arms, anchor guys, and equipment attached thereto less than 22 ft above the nearest track rail shall be located not less than 12 ft from the nearest track rail. See Rule 234H.

EXCEPTION 1: A clearance of not less than 7 ft may be allowed where the supporting structure is not the controlling obstruction, provided sufficient space for a driveway is left where cars are loaded or unloaded.

EXCEPTION 2: Supports for overhead trolley contact conductors may be located as near their own track rail as conditions require. If very close, however, permanent screens on cars will be necessary to protect passengers.

EXCEPTION 3: Where necessary to provide safe operating conditions which require an uninterrupted view of signals, signs, etc. along tracks, the parties concerned shall cooperate in locating structures to provide the necessary clearance.

232. Vertical Clearance of Wires, Conductors, Cables, and Live Parts of Equipment Above Ground, Rails, or Water

The vertical clearance of all wires, conductors, cables, and live parts of equipment above ground in generally accessible places, or above the top of the rails or water, shall not be less than the following.

A. Basic Clearances for Wires, Conductors, and Cables

The clearances in Table 232-1 apply under the following conditions:

1. Conductor temperature of 60°F, no wind, with final unloaded sag in the wire, conductors, or cables, or with initial unloaded sag in cases where these facilities are maintained approximately at initial unloaded sags.

2. Span lengths not greater than the following:

Loading district	Span lengths (feet)
Heavy	[1]175
Medium	[1]250
Light	350

[1]150 feet in heavy-loading district and 225 feet in medium-loading district for three-strand conductors, each wire of which is 0.09 in or less in diameter.

B. Additional Clearances for Wires, Conductors and Cables

Greater clearances than specified in Table 232-1 (Rule 232A) shall be provided where required by Rules 232B1 or 232B2. Increases are cumulative where more than one apply.

EXCEPTION 1: Additional clearances are not required for guys.

EXCEPTION 2: Additional clearances are not required for communication cables supported on messengers and communication wires which do not overhang the traveled way, but run along and within the limits of public highways or other public rights-of-way for traffic.

Table 232-1. Minimum Vertical Clearance of Wires, Conductors, and Cables Above Ground, Rails, or Water

(Voltages are phase to ground for effectively grounded circuits and those other circuits where all ground faults are cleared by promptly de-energizing the faulted section, both initially and following subsequent breaker operations. See the definition section for voltages of other systems.)

Nature of surface underneath wires, conductors, or cables	Communication conductors and cables, guys, messengers, surge protection wires, neutral conductors meeting Rule 230E1, and supply cables meeting Rule 230 C1 (11) (ft)	Supply line conductors, street lighting conductors, and service drops			Trolley and electrified railroad contact conductors and associated span or messenger wires (1)	
		Open supply line conductors 0 to 750 V. Supply cables of all voltages meeting Rule 230C2 or 230C3 (ft)	Open supply line conductors 750 V to 15 kV (ft)	15 to 50 kV (ft)	0 to 750 V to ground (ft)	750 V to 50 kV to ground (ft)
		Where wires, conductors, or cables cross over or overhang				
1. Track rails of railroads (except electrified railroads using over-head trolley conductors) (2)(16)(20)	(3)(15) 27	(3) 27	(3) 28	30	(4) 22	(4) 22
2. Roads, streets, alleys, nonresidential driveways, parking lots, and other areas subject to truck traffic (21)(22)	(6)(13)(23) 18	18	20	22	(5) 18	(5) 20
3. Residential driveways; commercial areas not subject to truck traffic (21) (22)	10	(8) 15	20	22	(5) 18	(5) 20

	(1)	(2)	(3)	(4)	(5)	(6)
4. Other land traversed by vehicles such as cultivated, grazing, forest, orchard, etc	18	18	20	22	—	—
5. Spaces or ways accessible to pedestrians only ⑨	⑦ 15	⑧⑭15	15	17	16	18
6. Water areas not suitable for sailboating or where sailboating is prohibited ⑲	15	15	17	17	—	—
7. Water areas suitable for sailboating including lakes, ponds, reservoirs, tidal waters, rivers, streams, and canals with an unobstructed surface area of: ⑰ ⑱ ⑲						
(a) Less than 20 acres	18	18	20	22	—	—
(b) 20 to 200 acres		26	28	30	—	—
(c) 200 to 2000 acres		32	34	36	—	—
(d) Over 2000 acres		38	40	42	—	—
8. Public or private land and water areas posted for rigging or launching sailboats	Clearance above ground shall be 5 ft greater than in 7 above, for the type of water areas served by the launching site					
Where wires, conductors, or cables run along and within the limits of highways or other road rights-of-way but do not overhang the roadway						
9. Roads, streets, or alleys in urban districts	⑬㉓18	18	20	22	⑤18	⑤20
10. Roads in rural districts where it is unlikely that vehicles will be crossing under the line	⑩⑫14	⑩15	18	20	⑤18	⑤20

(Footnotes for Table 232-1 on pages 144-146)

① Where subways, tunnels, or bridges require it, less clearances above ground or rails than required by Table 232-1 may be used locally. The trolley and electrified railroad contact conductor should be graded very gradually from the regular construction down to the reduced elevation.

② For wire, conductors, or cables crossing over mine, logging, and similar railways which handle only cars lower than standard freight cars, the clearance may be reduced by an amount equal to the difference in height between the highest loaded car handled and 20 ft, but the clearances shall not be reduced below that required for street crossings.

③ These clearances may be reduced to 25 ft where paralleled by trolley-contact conductor on the same street or highway.

④ In communities where 21 ft has been established, this clearance may be continued if carefully maintained. The elevation of the contact conductor should be the same in the crossing and next adjacent spans. (See Rule 289D2 for conditions which must be met where uniform height above rail is impractical.)

⑤ In communities where 16 ft has been established for trolley and electrified railroad contact conductors 0 to 750 V to ground, or 18 ft for trolley and electrified railroad contact conductors exceeding 750 V, or

where local conditions make it impractical to obtain the clearance given in the table, these reduced clearances may be used if carefully maintained.

⑥ If a communication service drop or a guy which is effectively grounded or is insulated against the highest voltage to which it is exposed, up to 8.7 kV, crosses residential streets and roads, the clearance may be reduced to 16 ft at the side of the traveled way provided the clearance at the center of the traveled way is at least 18 ft. This reduction in clearance does not apply to arterial streets and highways which are primarily for through traffic, usually on a continuous route.

⑦ This clearance may be reduced to the following values:

		feet
(a)	For insulated communication conductors and communication cables	8
(b)	For conductors of other communication circuits	10
(c)	For guys	8
(d)	For supply cables meeting Rule 230C1	10

Footnotes for Table 232-1 Continued on pages 145-146

⑧This clearance may be reduced to the following values:

                                     feet

(a) Supply conductors limited to 300 V to ground if more than 25 ft measured in any direction from a swimming pool, swimming area, or diving platform   12

(b) Drip loops of supply conductors limited to 150 V to ground and meeting Rules 230C2 or 230C3 and located at the electric service entrance to buildings.   10

(d) Supply conductors limited to 300 V to ground   12

(e) Guys   8

⑨Spaces and ways accessible to pedestrians only are areas where vehicular traffic is not normally encountered or not reasonably anticipated.

⑩Where a supply or communication line along a road is located relative to fences, ditches, embankments, etc, so that the ground under the line would not be expected to be traveled except by pedestrians, this clearance may be reduced to the following values:

                                     feet

(a) Insulated communication conductor and communication cables   8

(b) Conductors of other communication circuits   10

(c) Supply cables of any voltage meeting Rule 230C¹ and supply cables limited to 150 V to ground meeting Rules 230C2 or 230C3   10

⑪No clearance from ground is required for anchor guys not crossing track rails, streets, driveways, roads, or pathways.

⑫ This clearance may be reduced to 13 ft for communication conductors.

⑬Where communication wires or cables or supply cables meeting Rule 230C1 cross over or run along alleys, driveways, or parking lots, this clearance may be reduced to 15 ft for spans limited to 150 ft.

⑭Where supply circuits of 600 V or less, with transmitted power of 5000 W or less, are run along fenced (or otherwise guarded) private rights-of-way in accordance with the provisions specified in Rule 220B2, this clearance may be reduced to 10 ft.

⑮The value may be reduced to 25 ft for guys, for cables carried on messengers, and for supply cables meeting Rule 230C1. This value may be reduced to 25 ft for conductors effectively grounded throughout their length and associated with supply circuits of 0 to 22 kV, only if such conductors are stranded, are of corrosion-resistant material, and conform to the strength and tension requirements for messengers given in Rule 261I.

(Continued on page 146)

⑯ Adjacent to tunnels and overhead bridges which restrict the height of loaded rail cars to less than 20 ft, these clearances may be reduced by the difference between the highest loaded rail car handled and 20 ft, if mutually agreed to by the parties at interest.

⑰ For controlled impoundments, the surface area and corresponding clearances shall be based upon the design high water level. For other waters, the surface area shall be that enclosed by its annual high water mark, and clearances shall be based on the normal flood level. The clearance over rivers, streams, and canals shall be based upon the largest surface area of any 1 mi long segment which includes the crossing.. The clearance over a canal, river, or stream normally used to provide access for sailboats to a larger body of water shall be the same as that required for the larger body of water.

⑱ Where an overwater obstruction restricts vessel height to less than the following:

For a surface area in acres of	A reference vessel height in feet of
less than 20	16
20 to 200	24
200 to 2000	30
over 2000	36

the required clearance may be reduced by the difference between the reference vessel height given above and the overwater obstruction height, except that the reduced clearance shall not be less than that required for the surface area on the line crossing side of the obstruction.

⑲ Where the US Army Corps of Engineers, or the State, or a surrogate thereof has issued a crossing permit, clearances of that permit shall govern.

⑳ See Rule 234H for the required horizontal and diagonal clearances to rail cars.

㉑ These clearances do not allow for the future road resurfacing.

㉒ For the purpose of this rule, trucks are defined as any vehicle exceeding 8 ft in height.

㉓ For communications cables supported on a messenger, and with span lengths not exceeding 150 ft, clearance may be reduced to 17 ft above or along local streets or roads. This reduction does not apply for arterial streets or highways which are primarily for through traffic, usually on a continuous route.

1. **Voltages Exceeding 50 Kilovolts**

   a. For voltages between 50 and 470 kilovolts, the clearance specified in Table 232-1 (Rule 232A) shall be increased at the rate of 0.4 in per kilovolt in excess of 50 kilovolts. For voltages exceeding 470 kV, the clearance shall be determined by the alternate method given by Rule 232D. All clearances for lines over 50 kV shall be based on the maximum operating voltage.

      EXCEPTION: For voltages exceeding 98 kV alternating current to ground or 139 kV direct current to ground, clearances less than those required above are permitted for systems with known maximum switching surge factors (see Rule 232D).

   b. The additional clearance for voltages exceeding 50 kV specified in Rule 232B1a shall be increased 3% for each 1000 ft in excess of 3300 ft above mean sea level.

   c. For voltages exceeding 98 kV alternating current to ground, or 139 kV direct current to ground, either the clearances shall be increased or the electric field, or the effects thereof, shall be reduced by other means, as required, to limit the current due to electrostatic effects to 5.0 milliamperes, rms, if the largest anticipated truck, vehicle, or equipment under the line were shortcircuited to ground. For this determination, the conductors shall be at a final unloaded sag at 120 °F.

2. **Sag Increase**

   a. No additional clearance is required for trolley and electrified railroad contact conductors.

   b. No additional clearance is required where span lengths are less than those listed in Rule 232A2 and the maximum conductor temperature for which the supply line is designed to operate is 120°F or less.

   c. Where supply lines are designed to operate at or below a conductor temperature of 120°F and spans are longer than specified in Rule 232A2, the minimum clearance at midspan shall be increased by the following amounts.

      (1) General

          For spans exceeding the limits specified in Rule 232A2, the clearance specified in Table 232-1 shall be increased by 0.1 foot for each 10 feet of the excess of span length over such limits. See Rule 232B2c(3).

(2) Railroad Crossings

For spans exceeding the limits specified in Rule 232A2, the clearance specified in Table 232-1 shall be increased by the following amounts for each 10 feet by which the crossing span length exceeds such limits. See Rule 232B2c(3).

| Loading district | Amount of increase per 10 feet | |
	Large conductors (feet)	[1] Small conductors (feet)
Heavy and medium	0.15	0.30
Light	.10	.15

[1] A small conductor is a conductor having an over-all diameter of metallic material equal to or less than the following values:

| Material | Outside diameter of conductor | |
	Solid (inches)	Stranded (inches)
All copper	0.160	0.250
Other than all copper	.250	.275

(3) Limits

The maximum additional clearance need not exceed the arithmetic difference between final unloaded sag at a conductor temperature of $60°F$, no wind, and final sag at the following conductor temperature and condition, whichever difference is greater, computed for the crossing span.

(a) $32°F$, no wind, with radial thickness of ice, if any, specified in Rule 250B for the loading district concerned.

(b) $120°F$, no wind.

d. Where supply lines are designed to operate at conductor temperature above 120°F regardless of span length, the minimum clearance at midspan specified in Rule 232A and Rule 232B1 shall be increased by the difference between final unloaded sag at a conductor temperature of 60°F, no wind, and final sag at the following conductor temperature and condition, whichever difference is greater, computed for the crossing span.

(1) 32°F, no wind, with radial thickness of ice, if any, specified in Rule 250B for the loading district concerned.

(2) The maximum conductor temperature for which the supply line is designed to operate, with no horizontal displacement.

e. Where minimum clearance is not at midspan, the additional clearances specified in Rules 232B2c and 232B2d may be reduced by multiplying by the following factors:

Distance from nearer support of crossing span to point of crossing in percentage of crossing span length	Factors[1]
5	0.19
10	0.36
15	0.51
20	0.64
25	0.75
30	0.84
35	0.91
40	0.96
45	0.99
50	1.00

[1] Interpolate for intermediate values.

In applying this rule, the "point of crossing" is the location under the conductors of any topographical feature which is the determinant of the clearance.

Table 232-2. Minimum Vertical Clearance of Rigid Live Parts Above Ground

(Voltages are phase to ground for effectively grounded circuits and those other circuits where all ground faults are cleared by promptly de-energizing the faulted section, both initially and following subsequent breaker operations. See the definition section for voltages of other systems.)

Nature of surface below live parts	0 to 750 V (ft)	750 V to 15 kV (ft)	15 to 50 kV (ft)
1. Where live parts overhang:			
a. Roads, streets, alleys; nonresidential driveways; parking lots and other areas subject to truck traffic ④⑤	16	18	20
b. Residential driveways; commercial areas not subject to truck traffic ④⑤	①13	18	20
c. Other land traversed by vehicles such as cultivated land, grazing land, forest, orchard, etc.	16	18	20
d. Spaces and ways accessible to pedestrians only.⑥	①③④13	13	15
2. Where live parts are along and within the limits of highways or other road rights-of-way but do not overhang the roadway:			
a. Roads, streets, and alleys	②16	18	20
b. Roads in rural districts where it is unlikely that vehicles will be crossing under the line	②13	16	18

①This clearance may be reduced to the following values:

	feet
(a) Live parts limited to 300 V to ground	12
(b) Live parts limited to 150 V to ground and short lengths of supply cables meeting Rule 230C2 or 230C3 and located at the electric service entrance to building	10

② Where a supply line along a road is limited to 300 V to ground and is located relative to fences, ditches, embankments, etc, so that the ground under the line would not be expected to be traveled except by pedestrians, this clearance may be reduced to 12 ft.

(Footnotes continued on page 151)

C. Clearance to Live Parts of Equipment Mounted on Structures

1. Basic Clearances

The vertical clearance above ground for unguarded live parts such as potheads, transformer bushings, surge arresters, and short lengths of supply conductors connected thereto, which are not subject to variation in sag, shall be as shown in Table 232-2.

2. Additional Clearances for Voltages Exceeding 50 Kilovolts

a. For voltages between 50 and 470 kilovolts, the clearance specified in Table 232-2 (Rule 232C1) shall be increased at the rate of 0.4 in per kilovolt in excess of 50 kV. For voltages exceeding 470 kV, the clearances shall be determined by the alternate method given by Rule 232D. All clearances for lines over 50 kV shall be based on the maximum operating voltage.

EXCEPTION: For voltages exceeding 98 kV alternating current to ground or 139 kV direct current to ground, clearances less than those required above are permitted for systems with known maximum switching surge factors. (See Rule 232D.)

b. The additional clearance for voltages exceeding 50 kV specified in Rule 232C2a shall be increased 3% for each 1000 ft in excess of 3300 ft above mean sea level.

Footnotes for Table 232-2. *(Continued)*

③ Where supply circuits of 600 V or less, with transmitted power of 5000 W or less, are run along fenced (or otherwise guarded) private rights-of-way in accordance with the provisions specified in Rule 220B2, this clearance may be reduced to 10 ft.

④ For the purpose of this rule, trucks are defined as any vehicle exceeding 8 ft in height.

⑤ These clearances do not allow for future road resurfacing.

⑥ Spaces and ways accessible to pedestrians only are areas where vehicular traffic is not normally encountered or not reasonably anticipated.

Table 232-3.   Reference Heights

Nature of surface underneath lines	Feet
a. Track rails of railroads (except electrified railroads using overhead trolley conductors) ①	22
b. Streets, alleys, roads, driveways, and parking lots	14
c. Spaces and ways accessible to pedestrians only ②	9
d. Other land, such as cultivated, grazing, forest or orchard, which is traversed by vehicles	14
e. Water areas not suitable for sailboating or where sailboating is prohibited	14
f. Water areas suitable for sailboating including lakes, ponds, reservoirs, tidal waters, rivers, streams, and canals with unobstructed surface area ③④	
(1) less than 20 acres	18
(2) 20 to 200 acres	26
(3) 200 to 2000 acres	32
(4) over 2000 acres	38
g. In public or private land and water areas posted for rigging or launching sailboats, the reference height shall be 5 feet greater than in f above, for the type of water areas serviced by the launching site	

① See Rule 234H for the required horizontal and diagonal clearances to rail cars.

② Spaces and ways accessible to pedestrians only are areas where vehicular traffic is not normally encountered or not reasonably anticipated.

③ For controlled impoundments, the surface area and corresponding clearances shall be based upon the design high water level. For other waters, the surface area shall be that enclosed by its annual high water mark, and clearances shall be based on the normal flood level. The clearance over rivers, streams, and canals shall be based upon the largest surface area of any one-mile-long segment which includes the crossing. The clearance over a canal or similar waterway providing access for sailboats to a larger body of water shall be the same as that required for the larger body of water.

(Footnotes continued on page 153)

c. For voltages exceeding 98 kV alternating current to ground, or 139 kV direct current to ground either the clearances shall be increased or the electric field, or the effects thereof, shall be reduced by other means, as required, to limit the current due to electrostatic effects to 5.0 milliamperes, rms, if the largest anticipated truck, vehicle, or equipment under the line were short-circuited to ground.

D. **Alternate Clearances for Voltages Exceeding 98 Kilovolts Alternating Current to Ground or 139 Kilovolts Direct Current to Ground**

The clearances specified in Rules 232A, 232B, and 232C may be reduced for circuits with known switching surge factors but shall not be less than the values computed by adding the reference height to the electrical component of clearance.

1. Sag Conditions of Line Conductors

Minimum vertical clearances shall be maintained under the following conductor temperatures and conditions:

a. 32 °F no wind, with radial thickness of ice specified in Rule 250B for the loading district concerned.

b. 120 °F, no wind.

c. Maximum conductor temperature, for which the line is designed to operate, if greater than 120 °F, with no horizontal displacement.

2. Reference heights are shown in Table 232-3.

---

④ Where an overwater obstruction restricts vessel height to less than the following:

For a surface of	A reference vessel height of
(1) less than 20 acres	16 feet
(2) 20 to 200 acres	24 feet
(3) 200 to 2000 acres	30 feet
(4) over 2000 acres	36 feet

The required clearance may be reduced by the difference between the reference vessel height given above and the overwater obstruction height, except that the reduced clearance shall not be less than that required for the surface area on the line crossing side of the obstruction.

3. Electrical Component of Clearance
   a. The clearance computed by the following equation and listed in Table 232-4 shall be added to the reference heights specified in Table 232-3:

$$D = 3.28 \left[\frac{V \cdot (PU) \cdot a}{500\ K} \right]^{1.667} bc \qquad \text{(feet)}$$

where
     V maximum alternating current crest operating voltage to ground or maximum direct current operating voltage to ground in kilovolts;

     PU maximum switching surge factor expressed in per-unit peak voltage to ground and defined as a switching surge level for circuit breakers corresponding to 98 percent probability that the maximum switching surge generated per breaker operation does not exceed this surge level, or the maximum anticipated switching surge level generated by other means, whichever is greater;

     a = 1.15, the allowance for three standard deviations;

     b = 1.03, the allowance for nonstandard atmospheric conditions;

     c = 1.2, the margin of safety;

     K = 1.15, the configuration factor for conductor-to-plane gap.

   b. The value of *D* shall be increased 3 percent for each 1000 feet in excess of 1500 feet above mean sea level.

   c. Either the clearances shall be increased or the electric field, or the effects thereof, shall be reduced by other means, as required, to limit the current due to electrostatic effects to 5.0 milliamperes, rms, if the largest anticipated truck, vehicle, or equipment under the line were shortcircuited to ground. For this determination, the conductors shall be at a final unloaded sag at 120 °F.

4. Limit

The clearances derived from Rules 232D2 and 232D3 shall be not less than the clearances given in Tables 232-1 or 232-2 computed for 98 kilovolts alternating current to ground in accordance with Rules 232B1 or 232C2, respectively.

Table 232-4. Electrical Component of Clearance Above Ground or Rail in Rule 232D3a

(Add 3% for each 1000 ft in excess of 1500 ft above mean sea level. Increase clearance to limit electrostatic effects in accordance with Rule 232D3c.)

Maximum operating voltage phase-to-phase (kV)	Switching surge factor (per unit)	Switching surge (kV)	Electrical component of clearance (ft)
242	4.5 or less	839 or less	① 9.6
362	2.8 or less	839 or less	① 9.6
550	1.9 or less	839 or less	① 9.6
	2.0	898	10.8
	2.2	988	12.7
	2.4	1079	14.6
	2.6	1168	16.7
800	1.6	1045	13.9
	1.8	1176	16.9
	2.0	1306	20.1
	2.1 or more	1372 or more	②21.8

① Limited by Rule 232D4.
② Limited by Rules 232A and 232B.

233. Clearances Between Wires, Conductors, and Cables Carried on Different Supporting Structures

A. General

Crossings should be made on a common supporting structure, where practical. In other cases, the clearance in any direction between crossing or adjacent wires, conductors, or cables carried on different supporting structures shall not be less than the horizontal clearance required by Rule 233B or the vertical clearance required by Rule 233C, as applicable.

The applicable clearance, horizontal or vertical, required between wires, conductors, or cables shall be determined by a *clearance envelope* applied at the points on the relevant segments of the *conductor movement envelopes* at the location where the two conductors would be the closest together, as shown in Fig 233-1. The *conductor movement envelopes* shall be determined for each conductor involved in accordance with Rule 233A1. The *clearance envelope* shall be determined in accordance with Rule 233A2.

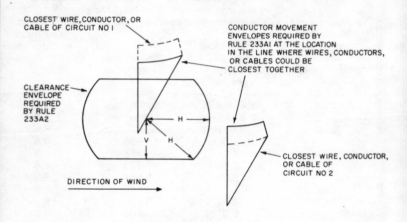

Fig 233-1
Use of Clearance Envelope and Conductor Movement
Envelopes to Determine Applicable Clearance

1. Conductor Movement Envelope
    a. Development
        The conductor movement envelope shall be developed
        from the locus of the most displaced conductor
        positions shown in Fig 233-2. The conductor posi-
        tions A—E which define the conductor movement
        envelope include the effects of the basic conditions
        shown in Fig 233-2 and the sag increases specified in
        Rule 233A1b as applicable.

Fig 233-2.
Conductor Movement Envelope

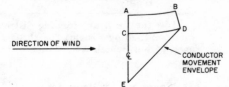

Point	Conductor Temperature	Sag	Ice Loading	Wind Displacement[1]
A	60° F	initial	none	none
B	60° F	initial	none	6 lb per sq ft[2]
C	60° F	final	none	none
D	60° F	final	none	6 lb per sq ft[2]
E_1[3][4]	The greater of 120° F or maximum operating	final	none	none
E_2[3][4]	32° F	final	as applicable	none

[1] The direction of the wind shall be that which produces the minimum separation. The displacement of the wire, conductors or cables includes the deflection of suspension insulators and flexible structures.

[2] Wind loading may be reduced to 4 lb per sq ft in areas sheltered by buildings, terrain, or other obstacles.

[3] If no sag increase is required by Rule 233A1b, point E = point C.

[4] Line D—E shall be considered to be straight unless the actual concavity characteristics are known.

b. Sag Increase
   (1) No sag increase is required for trolley and electrified railroad contact conductors.
   (2) No sag increase is required where span lengths are less than those listed below and the maximum conductor temperature for which the supply line is designed to operate is 120°F or less.

Loading district	Span lengths (feet)
Heavy	[1]175
Medium	[1]250
Light	350

[1]150 feet in heavy-loading district and 225 feet in medium-loading district for three-strand conductors, each wire of which is 0.09 inch or less in diameter.

   (3) Where supply lines are designed to operate at or below a conductor temperature of 120°F and spans are longer than specified in Rule 233A1b(2), the sag at midspan shall be increased by the following:
      (a) Where crossing occurs at midspan in the upper conductor sag shall be increased by the following amounts for each 10 feet by which the crossing span length exceeds the limits specified in Rule 233A1b(2).

Loading district	Amount of increase per 10 feet	
	Large conductors (feet)	Small[1] conductors (feet)
Heavy and medium	0.15	0.30
Light	.10	.15

[1]A small conductor is a conductor having an overall diameter of metallic material equal to or less than the following values:

	Outside diameter of conductor	
	Solid (inches)	Stranded (inches)
All copper	0.160	0.250
Other than all copper	.250	.275

(b) Limits

The maximum additional sag need not exceed the arithmetic difference between final unloaded sag at a conductor temperature of 60°F, no wind, and final sag at the conductor temperature and condition (i) or (ii) below, whichever difference is greater, computed for the crossing span.

(i) 32°F, no wind, with radial thickness of ice, if any, specified in Rule 250B for the loading district concerned.

(ii) 120°F, no wind.

(4) Where upper conductors are designed to operate at a conductor temperature above 120°F, the minimum sag at midspan specified in Rule 233A1a and Rule 233A1b(2) shall be increased by the difference between final unloaded sag at a conductor temperature of 60°F, no wind, and final sag at the following conductor temperature and condition, whichever difference is greater, computed for the crossing span.

(a) 32°F, no wind, with radial thickness of ice, if any, specified in Rule 250B for the loading district concerned.

(b) The maximum conductor temperature for which the supply line conductor is designed to operate, with no horizontal displacement.

(5) Where crossing is not at midspan of the upper conductor and under conditions where the upper span exceeds those specified in Rule 233A1b(2), the additional sag may be reduced by multiplying the additional sag determined by Rules 233A1b(3) and 233A1b(4) by the following factors:

159

Distance from nearest support of crossing span to point of crossing in percentage of crossing span length	Factors[1]
5	0.19
10	0.36
15	0.51
20	0.64
25	0.75
30	0.84
35	0.91
40	0.96
45	0.99
50	1.00

[1] Interpolate for intermediate values.

2. **Clearance Envelope**
   The clearance envelope shown in Fig 233-3 shall be determined by the horizontal clearance (H) required by Rule 233B and the vertical clearance (V) required by Rule 233C.

Fig 233-3.
Clearance Envelope

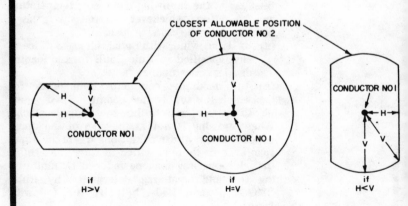

CLOSEST ALLOWABLE POSITION OF CONDUCTOR NO 2

CONDUCTOR NO I

if H>V

CONDUCTOR NO I

if H=V

CONDUCTOR NO I

if H<V

B. Horizontal Clearance

1. **Basic Clearance Requirements**

   The horizontal clearance between crossing or adjacent wires, conductors or cables carried on different supporting structures shall not be less than 5 ft. For voltages between the wires, conductors, or cables exceeding 129 kV, additional clearance of 0.4 in per kV over 129 kV shall be provided.

   EXCEPTION: The horizontal clearance between anchor guys of different supporting structures may be reduced to 6 in.

2. **Alternate Clearances for Voltages Exceeding 98 kV Alternating Current to Ground or 139 kV Direct Current to Ground**

   The clearances specified in Rule 233B1 may be reduced for circuits with known switching surge factors but shall not be less than the clearances derived from the computations required in Rules 235B3a and 235B3b.

C. Vertical Clearance

1. **Basic Clearance**

   The vertical clearance between any crossing or adjacent wires, conductors, or cables carried on different supporting structures shall not be less than those shown in Table 233-1.

2. **Voltages Exceeding 50 Kilovolts**

   a. The clearance given in Table 233-1 shall be increased by the sum of the following: For the upper level conductors between 50 and 470 kilovolts, the clearance shall be increased at the rate of 0.4 inch per kilovolt in excess of 50 kilovolts. For the lower level conductors exceeding 50 kilovolts, the additional clearance shall be computed at the same rate. For voltages exceeding 470 kilovolts, the clearance shall be determined by the alternate method given in Rule 233C3. The additional clearance shall be computed using the maximum operating voltage if above 50 kilovolts and nominal voltage if below 50 kilovolts.

   EXCEPTION: For voltages exceeding 98 kilovolts alternating current to ground or 139 kilovolts direct current to ground, clearances less than those required above are permitted for systems with known switching surge factors. (See Rule 233C3.)

Table 233-1. Vertical Clearances of Wires, Conductors, and Cables Carried on Different Supporting Structures

(Voltages are phase to ground for effectively grounded circuits and those other circuits where all ground faults are cleared by promptly de-energizing the faulted section, both initially and following subsequent breaker operations. See the definition section for voltages of other systems.)

(The insertion of a given clearance in *italics* indicates that in general, the lines operating at the voltage named above this clearance should not cross over the lines at the voltage to the left of the clearance in italics.)

Upper level → Lower level ↓	Communications conductors, cables, and messengers (ft)	Supply cables and messengers meeting Rule 230C1 (including service drops over 750 V) (ft)	Open supply conductors, 0 to 750 V; Supply cables meeting Rule 230C2 or 230C3		Open supply conductors and service drops		Guys, span wires, neutral conductors meeting Rule 230E1, and lightning protection wires (ft)
			Line conductors (ft)	Service drops (ft)	750 V to 8.7 kV (ft)	8.7 to 50 kV (ft)	
Communications conductors, cables, and messengers	[2]2	2	4	[6]2	[5]4	6	2
Supply cables and messengers meeting Rule 230C1 (including service drops over 750 V)	[6]2	[6]2	[6]2	[6]2	[6]2	4	[6]2
Supply cables and messengers of any voltage meeting Rule 230C2 or 230C3 (including service drops over 750 V)	4	4	[6]2	[6]2	[6]2	4	2

Open supply service drops (0 to 750 V)	4	4	⑥2	⑥2	4	4	2
Open supply conductors (0 to 750 V)	4	4	⑥2	⑥2	⑥2	4	2
750 V to 8.7 kV	4	4	⑥2	4	⑥2	4	4
8.7 to 50 kV	6	6	4	6	4	4	4
Trolley and electrified railroad contact conductors and associated span and messenger wires	③4	③4	③④4	③4	③4	6	③4
Guys⑦, span wires, neutral conductors meeting Rule 230E1, and lightning protection wires	②2	②2	⑥2	⑥2	6	4	①②2

①This clearance may be reduced where both guys are electrically interconnected.

②The clearance of communication conductors and their guy, span, and messenger wires from each other in locations where no other classes of conductors are involved may be reduced by mutual consent of the regulatory body having jurisdiction, except for fire-alarm conductors and conductors used in the operation of the railroads, or where one set of conductors is for public use and the other used in the operation of supply systems.

③Trolley and electrified railroad contact conductors of more than 750 V should have at least 6 ft clearance. This clearance should also be provided over lower voltage trolley and electrified railroad contact conductors unless the crossover conductors are beyond reach of a trolley pole leaving the trolley-contact conductor or are suitably protected against damage from trolley poles leaving the trolley-contact conductor.

④Trolley and electrified railroad feeders are exempt from this clearance requirement for contact conductors if they are of the same nominal voltage and of the same system.

(Footnotes for Table 233-1 continued on page 164)

Footnotes for Table 233-1. (Continued)

⑤ This clearance shall be increased to 6 ft where the supply wires cross over a communication line within 6 ft horizontally of a communication structure.

⑥ Where a 2 ft clearance is required at 60 °F, and where conditions are such that the sag in the upper conductor would increase more than 1.5 ft at the crossing point under any condition of sag stated in Rule 233A 1b and c, and the 2 ft clearance shall be increased by the amount of sag increase less 1.5 ft.

⑦ These clearances may be reduced by not more than 25% to a guy insulator, provided that full clearance is maintained to its metallic end fittings and the guy wires. The clearance to an insulated section of a guy between two insulators may be reduced by not more than 25% provided that full clearance is maintained to the uninsulated portion of the guy.

b. The additional clearance for voltages in excess of 50 kilovolts specified in Rule 233C2a shall be increased 3 percent for each 1000 feet in excess of 3300 feet above mean sea level.

3. Alternate Clearances for Voltages Exceeding 98 Kilovolts Alternating Current to Ground or 139 Kilovolts Direct Current to Ground
The clearances specified in Rules 233C1 and 233C2 may be reduced where the higher voltage circuit has a known switching surge factor. The clearances shall not be less than the values computed by adding the reference heights to the electrical component of clearance.

a. Reference Heights

Reference height	Feet
(1) Supply lines	0
(2) Communication lines	2

b. Electrical Component of Clearance
(1) The alternate clearance is computed by the following equation and listed in Table 233-2:

$$D = 3.28 \left[\frac{[V_H \cdot (PU) + V_L]\, a}{500\, K} \right]^{1.667} bc \qquad \text{(feet)}$$

Table 233-2. Clearance Between Supply Wires, Conductors, and Cables in Rule 233C3b(1)
(Add 3 percent for each 1000 ft in excess of 1500 ft above mean sea level.)

Higher voltage circuit		Lower voltage circuit						
Maximum operating voltage phase to phase (kV)	Switching surge factor (per unit)	Maximum operating voltage, phase to phase (kV)						
		121 (ft)	145 (ft)	169 (ft)	242 (ft)	362 (ft)	550 (ft)	800 (ft)
242	3.3 or less	① 7.0	① 7.0	① 7.0	① 7.0			
362	2.4	① 9.3	① 9.3	① 9.3	① 9.3	9.4		
	2.6	① 9.3	① 9.3	① 9.3	① 9.3	10.3		
	2.8	① 9.3	① 9.3	① 9.3	9.7	11.3		
	3.0	① 9.3	① 9.4	9.7	10.7	12.3		
550	1.8	① 13.0	① 13.0	① 13.0	① 13.0	① 13.0	13.6	
	2.0	① 13.0	① 13.0	① 13.0	① 13.0	① 13.0	15.3	
	2.2	① 13.0	① 13.0	① 13.0	① 13.0	14.1	17.0	
	2.4	① 13.0	① 13.0	① 13.0	14.0	15.8	18.8	
	2.6	② 13.6	② 14.1	14.5	15.6	17.5	20.7	
800	1.6	① 17.7	① 17.7	① 17.7	① 17.7	① 17.7	18.5	22.5
	1.8	① 17.7	① 17.7	① 17.7	① 17.7	① 17.7	20.9	25.4
	2.0	① 17.7	① 17.7	① 17.7	18.4	20.4	23.1	27.5
	2.2	② 18.4	② 18.9	② 19.4	② 20.8	② 23.1	② 26.7	② 30.8

① Limited by Rule 233C3c.
② Need not be greater than the values specified in Rules 233C1 and 233C2.

where

V_H higher voltage circuit maximum alternating current crest operating voltage to ground or maximum direct current operating voltage to ground in kilovolts;

V_L lower voltage circuit maximum alternating current crest operating voltage to ground or maximum direct current operating voltage to ground in kilovolts;

PU higher voltage circuit maximum switching surge factor expressed in per-unit peak voltage to ground and defined as a switching surge level for circuit breakers corresponding to 98 percent probability that the maximum switching surge generated per breaker operation does not exceed this surge level, or the maximum anticipated switching surge level generated by other means, whichever is greater;

a = 1.15, the allowance for three standard deviations;

b = 1.03, the allowance for nonstandard atmospheric conditions;

c = 1.2, the margin of safety;

K = 1.4, the configuration factor for conductor-to-conductor gap.

(2) The value of D calculated by Rule 233C3b(1) shall be increased 3 percent for each 1000 feet in excess of 1500 feet above mean sea level.

c.   Limit

The value of D shall not be less than the clearance required by Rules 233C1 and 233C2 with the lower voltage circuit at ground potential.

234. Clearance of Wires, Conductors, and Cables from Buildings, Bridges, Rail Cars, Swimming Pools, and Other Installations

A.   Application

The basic vertical and horizontal clearances specified in Rules 234.B, C, D, E, and H apply under the following conditions.

1.   Horizontal Clearance

Clearances shall be applied with the wire, conductor, or cable displaced from rest by a 6 pound per square foot wind at final sag at 60°F. This may be reduced to a 4 pound per square foot wind in areas sheltered by buildings, terrain, or other obstacles. The displacement of the wire, conductor, or cable shall include deflection of suspension insulators and flexible structures.

2. Vertical Clearance

    a. Conductor temperature of 60°F, no wind, with final unloaded sag in the wire, conductors, or cables, or with initial unloaded sag in cases where these facilities are maintained approximately at initial unloaded sags.

    b. Span lengths not greater than the following:

Loading district	Span lengths (feet)
Heavy	[1]175
Medium	[1]250
Light	350

[1]150 feet in heavy-loading district and 225 feet in medium-loading district for three-strand conductors, each wire of which is 0.09 inch or less in diameter.

3. Diagonal Clearance

The horizontal clearance governs above the roof level or top of an installation to the point where the diagonal equals the vertical clearance requirement. Similarly, the horizontal clearance governs above or below projections from buildings, signs, or other installations to the point where the diagonal equals the vertical clearance requirement. The 15 feet for roofs accessible to pedestrians agrees with Table 232-1 for spaces and ways accessible to pedestrians only. From this point the diagonal clearance shall equal the vertical clearance as shown in Figure 234-1. This rule should not be interpreted as restricting the installation of a trolley-contact conductor over the approximate center line of the track it serves.

B. Clearances of Wires, Conductor, and Cables from Other Supporting Structures

Wires, conductors, or cables of one line passing near a lighting support, traffic signal support, or a supporting structure of a second line, without being attached thereto, shall have clearance from any part of such structure not less than the following:

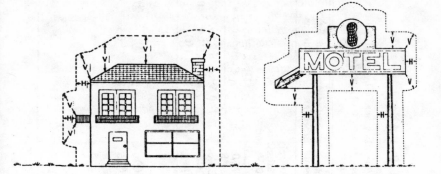

V = Minimum vertical clearance, measured
     either diagonally or vertically.

H = Minimum horizontal clearance.

Fig 234-1
Clearance Diagram for Building and
Other Structures

1. A minimum horizontal clearance of 5 feet for voltages up to 50 kilovolts.
2. A minimum vertical clearance of 6 feet for voltages below 15 kilovolts and a minimum vertical clearance of 7 feet for voltages between 15 and 50 kilovolts.

   EXCEPTION 1: Where the voltage does not exceed 300 V to ground and the cables meet the requirements of rule 230C1, 230C2 or 230C3, the vertical and horizontal clearances may be reduced to 4 feet, measured at 60 °F, without wind deflection.

   NOTE: Clearances of wires, conductors, and cables from adjacent line structure guy wires are given in Rule 233.

   EXCEPTION 2: The vertical clearances may be reduced by 2 ft if both of the following conditions are met:
   a. The wires, conductors, or cables above and the supporting structure of another line below are operated and maintained by the same utility.

Table 234-1. Clearance of Supply Wires, Conductors, and Cables Passing By But Not Attached To Building and Other Installations Except Bridges

(Voltages are phase to ground for effectively grounded circuits and those other circuits where all ground faults are cleared by promptly de-energizing the faulted section, both initially and following subsequent breaker operations. See the definitions section for voltages of other systems.)

Clearance of	Communication conductors and cables, guys, messengers, lightning protection wires, neutral conductors meeting Rule 230E1, supply cables of all voltages meeting Rule 230C1, and supply cables of 0 to 750 V meeting Rule 230C2 or 230C3 (ft)	Supply line conductors, street lighting conductors, and service drops			
		Open supply line conductors 0 to 750 V, and supply cables over 750 V meeting Rule 230C2 or 230C3 (ft)	Open supply line conductors		
			750 V to 8.7 kV (ft)	8.7 to 15 kV (ft)	15 to 50 kV (ft)
Buildings					
Horizontal					
To walls and projections	3	②①5	①②5	③8	③10
To unguarded windows	3	②①5	5	8	10
To balconies and areas accessible to pedestrians ④	3	5	5	8	10
Vertical					
Above or below roofs or projections not accessible to pedestrians ④	3	10	10	10	12

170

Above or below balconies and roofs accessible to pedestrians ④	8	⑥15	15	15	17
Above roofs accessible to truck traffic ⑦	18	18	20	20	22
Above roofs accessible to vehicles, but not subject to truck traffic ⑦	10	⑥15	20	20	22
Signs, chimneys, radio and television antennas, tanks, and other installations not classified as buildings or bridges ⑤					
Horizontal	3	①②5	①②5	③8	③10
Vertical above or below	3	①5	8	8	10

① Where building, sign, chimney, antenna, tank, or other installation does not require maintenance such as painting, washing, changing of sign letters, or other operation which would require persons to work or pass between supply conductors and structure, the clearance may be reduced to 3 ft.

② Where available space will not permit this value, the clearance may be reduced to the maximum practical clearance but the minimum clearance may not be less than 3 ft provided the conductors, including splices and taps, have covering which provides sufficient dielectric to prevent a short circuit in case of a momentary contact between the conductors and a grounded surface.

③ Where available space will not permit these values, the clearance may be reduced by 2 ft if the conductors, including splices and taps, have covering which provides sufficient dielectric to prevent a short circuit in case of a momentary contact between the conductors and a grounded surface.

④ A roof, balcony, or area is considered accessible to pedestrians if the means of access is through a doorway, ramp, stairway, or permanently mounted ladder.

⑤ The required clearances shall be increased to allow for the movement of motorized signs and other movable attachments to any installation covered by Rule 234C.

⑥ This clearance may be reduced to 12 ft for supply conductors limited to 300 V to ground if more than 25 ft measured in any direction from a swimming pool, swimming area, or diving platform.

⑦ For the purpose of this rule, trucks are defined as any vehicles exceeding 8 feet in height.

171

b. Employees do not work above the top of the supporting structure while the upper circuit is energized.

C. Clearances of Wires, Conductors, and Cables from Buildings; Signs, Chimneys, Radio and Television Antennas, Tanks Containing Nonflammables, and Other Installations Except Bridges

1. Ladder Space
Where buildings or other installations exceed three stories (or 50 feet) in height, overhead lines should be arranged where practical so that a clear space or zone at least 6 feet wide will be left either adjacent to the building or beginning not over 8 feet from the building, to facilitate the raising of ladders where necessary for fire fighting.

EXCEPTION: This requirement does not apply where it is the unvarying rule of the local fire departments to exclude the use of ladders in alleys or other restricted places which are generally occupied by supply conductors and cables.

2. Basic Clearances
Unguarded or accessible supply wires, conductors, or cables may be run either beside or over buildings or other installations and any projections therefrom. The vertical and horizontal clearances of such wires, conductors, or cables shall be not less than the values given in Table 234-1.

3. Guarding of Supply Conductors
Where the clearances set forth in Table 234-1 cannot be obtained, supply conductors shall be guarded.

NOTE: Metal-clad supply cables meeting Rule 230C1 are considered to be guarded within the meaning of this rule.

4. Supply Conductors Attached to Buildings
Where the permanent attachment of supply conductors of any class to building is necessary for an entrance, such conductors shall meet the following requirements:

a. Conductors of more than 300 volts to ground shall not be carried along or near the surface of the building unless they are guarded or made inaccessible.

b. Clearance of wires from building surface shall be not less than those required in Table 235-6 (Rule 235E1) for clearance of conductors from supports.

c. Service-drop conductors shall not be readily accessible and when not in excess of 600 volts they shall have a clearance of not less than 8 feet from the highest point of roofs over which they pass with the following exceptions:

EXCEPTION 1: Where the voltage between conductors does not exceed 300 volts and the roof is not readily accessible, the clearance may be not less than 3 feet. A roof is considered readily accessible if the means of access is through a doorway, ramp, stairway, or permanently mounted ladder.

EXCEPTION 2: Service-drop conductors of 300 volts or less which do not pass over other than a maximum of 4 feet of the overhang portion of the roof for the purpose of terminating at a (through-the-roof) service raceway or approved support, may be maintained at a minimum of 18 inches from any portion of the roof over which they pass.

5. **Communications Conductors Attached to Buildings**
Communications conductors and cables may be attached directly to buildings.

D. Clearances of Wires, Conductors, and Cables from Bridges

1. Basic Clearances
Supply wires, conductors, and cables which pass under, over, or near a bridge shall have basic vertical and horizontal clearances therefrom not less than given in Table 234-2.

EXCEPTION: This rule does not apply to guys, span wire, effectively grounded lightning protection wires, neutrals meeting Rule 230E1, and supply cables meeting Rule 230C1.

2. Guarding Trolley-Contact Conductors Located Under Bridges
a. Where Guarding is Required
Guarding is required where the trolley-contact conductor is located so that a trolley pole leaving the conductor can make simultaneous contact between it and the bridge structure.
b. Nature of Guarding
Guarding shall consist of a substantial inverted trough of nonconducting material located above the contact

Table 234-2. Clearance of Supply Conductors from Bridges

(Voltages are phase to ground for effectively grounded circuits and those others circuits where all ground faults are cleared by promptly de-energizing the faulted section, both initially and following breaker operations. See definition section for voltages of other systems.)

	Supply cables meeting Rule 230C2 or 230C3; Neutral conductors meeting Rule 230E1 (ft)	Open supply line conductors			
		0 to 750 V (ft)	750 V to 8.7 kV (ft)	8.7 to 15 kV (ft)	15 to 50 kV (ft)
Clearance over bridges①					
Attached③	3	3	3	5	8
Not attached	10	10	10	10	10
Clearance beside, under, or within bridge structure⑥					
Readily accessible portions of any bridge including wing, walls, and bridge attachments①					
Attached③	3	3	3	5	8
Not attached	5	5	5	8	10
Ordinarily inaccessible portions of bridges (other than brick, concrete, or masonry)② and from abutments②					
Attached③⑤	0.5	0.5	3	5	8
Not attached④⑤	3	3	3	8	10

① Where over traveled ways on or near bridges, the clearances of Rule 232 apply also.

② Bridge seats of steel bridges carried on masonry, brick, or concrete abutments which require frequent access for inspection shall be considered as readily accessible portions.

③ Clearance from supply conductors to supporting arms and brackets attached to bridges shall be the same as specified in Table 235-6 (Rule 235E1) if the supporting arms and brackets are owned, operated, or maintained by the same utility.

④ Conductors should have the clearances given in this row increased as much as is practical.

⑤ Where conductors passing under bridges are adequately guarded against contact by unauthorized persons and can be de-energized for maintenance of the bridge, clearances of the conductors from the bridge, at any point, may have the clearances specified in Table 235-6 for clearance from surfaces of support arms plus one-half the final unloaded sag of the conductor at that point.

⑥ Where the bridge has moving parts, such as a lift bridge, the required clearances shall be maintained throughout the full range of movement of the bridge or any attachment thereto.

175

conductor, or of other suitable means of preventing contact between the trolley support and the bridge structure.

E. Clearance of Wires, Conductors, or Cables Installed Over or Near Swimming Areas

1. Swimming Pools

    Where wires, conductors, or cables cross over a swimming pool or the surrounding area within 25 ft of the edge of the pool, the clearances in any direction shall be as shown in Fig 234-2. The values of A, B, and C are specified in Table 234-3.

    EXCEPTION: This rule does not apply to a pool fully enclosed by a solid or screened permanent structure.

2. Beaches and Waterways Restricted to Swimming

    Where rescue poles are used by lifeguards at supervised swimming beaches, the required vertical and horizontal clearances shall be as specified in Table 234-3. Where rescue poles are not used, the minimum clearances shall be as specified in Rule 232.

3. Waterways Subject to Water Skiing

    The minimum vertical clearance shall be the same as that specified in Rule 232.

Fig 234-2
Swimming Pool Clearances

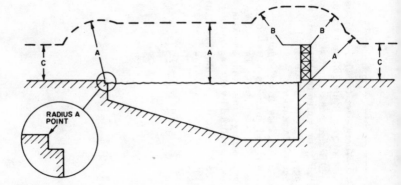

Table 234-3. Clearance of Supply Wires, Conductors, and Cables Passing Over or Near Swimming Areas

(Voltages are phase to ground for effectively grounded circuits and those other circuits where all ground faults are cleared by promptly de-energizing the faulted section, both initially and following subsequent breaker operations. See the definitions section for voltages of other systems.)

	Supply line conductors, street lighting conductors, and service drops				
	Supply cables, 0 to 750 V meeting Rules 230C2 or 230C3; Neutral conductors meeting Rule 230E1 (ft)	Open supply line conductors 0 to 750 V and supply cables over 750 V meeting Rules 230C2 or 230C3 (ft)	Open supply line conductors		
			750 V 8.7 kV (ft)	8.7 to 15 kV (ft)	15 to 50 kV (ft)
A: Clearance in any direction from the water level, edge of pool, base of diving platform, or anchored raft	18	25	25	25	27
B: Clearance in any direction to the diving platform or tower	14	16	16	16	18
C: Vertical clearance over adjacent land	Clearance shall be as required by Rule 232				

① A, B, and C are shown in Figure 234-2.

- F. Additional Clearance

    Greater clearances than the basic clearances specified in Rules 234B, C, D, and E shall be provided where the conditions exceed the basic conditions specified in Rule 234A. All increases are cumulative.

    1. Voltages Exceeding 50 Kilovolts

        The basic vertical and horizontal clearances specified in Rules 234B, C, D, and E shall be increased at the following rates:

        a. For voltages between 50 and 470 kilovolts, the clearances specified in Rules 234B, C, D, and E shall be increased at the rate of 0.4 inch per kilovolt in excess of 50 kilovolts. For voltages exceeding 470 kilovolts, the clearance shall be determined by the alternate method given by Rule 234G. All clearances for lines over 50 kilovolts shall be based on the maximum operating voltage.

            EXCEPTION: For voltages exceeding 98 kilovolts alternating current to ground or 139 kilovolts direct current to ground, clearances less than those required above are permitted for systems with known maximum switching surge factor. (See Rule 234G.)

        b. The additional clearance for voltages in excess of 50 kilovolts specified in Rule 234F1a shall be increased 3 percent for each 1000 feet in excess of 3300 feet above mean sea level.

        c. For voltages exceeding 98 kilovolts alternating current to ground, or 139 kilovolts direct current to ground, either the clearances shall be increased or the electric field, or the effects thereof, shall be reduced by other means, as required, to limit the current due to electrostatic effects to 5.0 milliamperes, rms, if an ungrounded, metal fence, building, sign, chimney, radio or television antenna, tank containing nonflammables or other installation, or any ungrounded metal attachments thereto were short-circuited to ground. For this determination, the conductor sag shall be at final unloaded sag at 120 °F.

    2. Sag Increase

        a. No additional clearance is required for trolley and electrified railroad contact conductors.

        b. No additional clearance is required where span lengths are less than those listed in Rule 234A2b and the

178

maximum conductor temperature for which the supply line is designed to operate is 120 °F or less.

c. Where supply lines are designed to operate at or below a conductor temperature of 120 °F and spans are longer than specified in Rule 234A2b, the minimum vertical clearance at midspan shall be increased by 0.1 foot for each 10 feet in excess of span length over such limits. The maximum additional clearance need not exceed the arithmetic difference between final unloaded sag at a conductor temperature of 60 °F, no wind, and final sag at the following conductor temperature and condition, whichever difference is greater, computed for the crossing span.

(1) 32°F, no wind, with radial thickness of ice, if any, specified in Rule 250B for the loading district concerned.

EXCEPTION: The additional clearances for ice loadings are not applicable to swimming pools (Rule 234E1).

(2) 120°F, no wind.

d. Where supply lines are designed to operate at conductor temperature above 120°F regardless of span length, the minimum vertical clearance at midspan specified in Rules 234B, C, D, E, and F1 shall be increased by the difference between final unloaded sag at a conductor temperature of 60°F, no wind, and final sag at the following conductor temperature and condition, whichever difference is greater, computed for the crossing span.

(1) 32°F, no wind, with radial thickness of ice, if any, specified in Rule 250B for the loading district concerned.

EXCEPTION: The additional clearances for ice loadings are not applicable to swimming pools (Rule 234E1).

(2) The maximum conductor temperature for which the supply line is designed to operate, with no horizontal displacement.

e. Where minimum clearance is not at midspan, the additional clearances specified in Rules 234F2c and 234F2d may be reduced by multiplying by the following factors:

Distance from nearer support of crossing span to point of crossing in percentage of crossing span length	Factors[1]
5	0.19
10	0.36
15	0.51
20	0.64
25	0.75
30	0.84
35	0.91
40	0.96
45	0.99
50	1.00

[1] Interpolate for intermediate values.

In applying the above rules, the "point of crossing" is the location of any topographical feature which is the determinant of the clearance.

G. **Alternate Clearances for Voltages Exceeding 98 Kilovolts Alternating Current to Ground or 139 Kilovolts Direct Current to Ground**
The clearances specified in Rules 234B, C, D, E, and F may be reduced for circuits with known switching surge factors but shall not be less than the values computed by adding the reference distance to the electrical component of clearance.
1. Sag Conditions
   a. Minimum vertical clearances shall be maintained under the following conductor temperatures and conditions:
      (1) 32°F, no wind, with radial thickness of ice specified in Rule 250B for the loading district concerned.
      (2) 120°F, no wind.
      (3) Maximum conductor temperature for which the line is designed to operate, if greater than 120°F.
   b. Horizontal and diagonal clearances shall be maintained under the conditions specified in Rules 234A1 and 234A3.

2.   Reference Distances

Reference distance	Horizontal (feet)	Vertical (feet)
a. Buildings	5	9
b. Signs, chimneys, radio and television antennas, tanks, and other installations not classified as bridges or buildings	5	9
c. Superstructure of bridges[1] , [2]	5	9
d. Supporting structures of another line	5	6
e. Dimension *A* of Figure 234-2	—	18
f. Dimension *B* of Figure 234-2	14	14

[1]Where overtraveled ways on or near bridges, the clearances of Rule 232 apply also.

[2]Where the bridge has moving parts, such as a lift bridge, the required clearances shall be maintained throughout the full range of movement of the bridge or any attachment thereto.

3.   Electrical Component of Clearance
   a. The clearance computed by the following equation and listed in Table 234-4 shall be added to the reference distance specified in Rule 234G2:

$$D = 3.28 \left[\frac{V \cdot (PU) \cdot a}{500\,K} \right]^{1.667} bc \quad \text{(feet)}$$

where
   V   maximum alternating current crest operating voltage to ground or maximum direct current operating voltage to ground in kilovolts;
   PU maximum switching surge factor expressed in per-unit peak voltage to ground and defined as a switching surge level for circuit breakers corresponding to 98 percent probability that the maximum switching surge generated per breaker operation does not exceed this surge level, or the maximum anticipated switching surge level generated by other means, whichever is greater;
   a = 1.15, the allowance for three standard deviations;

b = 1.03, the allowance for nonstandard atmospheric conditions;

c = margin of safety
1.2 for vertical clearances
1.0 for horizontal clearances

K = 1.15, the configuration factor for conductor-to-plane gap.

Table 234-4.   Electrical Component of Clearance to Buildings, Bridges, and Other Installations in Rule 234G3a
(Add 3 percent for each 1000 ft in excess of 1500 ft above mean sea level.)

Maximum operating voltage phase to phase (kV)	Switching surge factor (per unit)	Switching surge (kV)	Electrical component of clearances	
			V (ft)	H (ft)
242	2.0	395	2.7	2.3
	2.2	435	3.2	2.7
	2.4	474	3.7	3.1
	2.6	514	4.2	3.5
	2.8	553	4.8	4.0
	3.0	593	5.4	4.5
362	1.8	532	4.5	3.7
	2.0	591	5.4	4.5
	2.2	650	6.3	5.2
	2.4	709	7.3	6.1
	2.6	768	8.3	6.9
	2.8	828	9.4	7.8
	3.0	887	10.6	8.8
550	1.6	719	7.5	6.2
	1.8	808	9.1	7.6
	2.0	898	10.8	9.0
	2.2	988	12.7	10.6
	2.4	1079	14.6	12.2
	2.6	1168	16.7	13.9
800	1.6	1045	13.9	11.6
	1.8	1176	16.9	14.1
	2.0	1306	20.1	16.7
	2.2	1437	23.6	19.7
	2.4	1568	27.3	22.7

b. The value of D above shall be increased by 3 percent for each 1000 feet in excess of 1500 feet above mean sea level.

4. Limit
The clearances derived from Rules 234G2 and 234G3 shall not be less than the basic clearances of Rule 234B, Tables 234-1 and 234-2, computed for 98 kilovolts alternating current rms to ground by Rule 234F1.

H. Clearance to Rail Cars
Where overhead wires, conductors, or cables run along railroad tracks, the minimum clearance in any direction shall be as shown in Figure 234-3. The values of V and H are defined as follows:

V   minimum vertical clearance from the wire, conductor, or cable above the top of the rail as specified in Rule 232 minus 20 feet, the assumed height of the rail car;

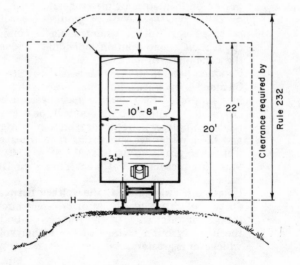

Fig 234-3
Rail Car Clearances

H   minimum horizontal clearance from the wire, conductor, or cable to the nearest rail, which is equal to the required vertical clearance above the rail minus 15 feet as computed by the lesser of the following:
1. Rule 232A and 232B1.
2. Rule 232D.

These clearances are computed for railroads handling standard rail cars as common carriers in interchange service with other railroads. Where wires, conductors, or cables run along mine, logging, and similar railways which handle only cars smaller than standard freight cars, the value of *H* may be reduced by one-half the difference between the width of a standard rail car (10 feet, 8 inches) and the width of the narrower car.

235. Clearance for Wires, Conductors, or Cables Carried on the Same Supporting Structure

A. Application of Rule
1. Multiconductor Wires or Cables
Cables, and duplex, triple, or paired conductors supported on insulators or messengers, meeting Rules 230C or 230D, whether single or grounded, are for the purposes of this rule considered single conductors even though they may contain individual conductors not of the same phase or polarity.
2. Conductors Supported by Messengers or Span Wires
Clearances between individual wires, conductors, or cables supported by the same messenger, or between any group and its supporting messenger, or between a trolley feeder, supply conductor, or communication conductor, and their respective supporting span wires, are not subject to the provisions of this rule.
3. Line Conductors of Different Phases on Different Circuits
Unless otherwise stated, the voltage between line conductors of different phases of different circuits shall be the vector difference of the voltages of both circuits or the line-to-ground voltage of the higher voltage circuit, whichever is greater.

B. Horizontal Clearance Between Line Conductors
1. Fixed Supports
Line conductors attached to fixed supports shall have horizontal clearances from each other not less than the

larger value required by either Rule 235B1a or Rule 235B1b for the situation concerned.

EXCEPTION 1: The pin spacing at buckarm construction may be reduced as specified in Rule 236F to provide climbing space.

EXCEPTION 2: Grades D and N need meet only the requirements of Rule 235B1a.

EXCEPTION 3: These clearances do not apply to cables meeting Rule 230C or covered conductors of the same circuit meeting Rule 230D.

EXCEPTION 4: For voltages to ground exceeding 98 kilovolts alternating current or 139 kilovolts direct current, clearances less than those required by *a* and *b* below are permitted for systems with known maximum switching surge factors. (See Rule 235B3.)

a. Minimum Horizontal Clearance Between Line Conductors of the Same or Different Circuits

Clearances shall be not less than given in Table 235-1.

b. Clearance According to Sags

The clearance at the supports of conductors of the same or different circuits of grade B or C shall in no case be less than the values given by the following formulas, at a conductor temperature of 60°F, at final unloaded sag, no wind. All voltages are between the two conductors for which the clearance is being determined except for railway feeders which are to ground. The requirements of Rule 235B1a apply if they give a greater separation than this rule.

EXCEPTION: No requirement is specified for clearance between conductors of the same circuit when rated above 50 kilovolts.

In the following, S is the apparent sag in inches of the conductor having the greater sag, and the clearance is in inches.

(1) For line conductors smaller than No 2 AWG:
clearance = 0.3 in per kilovolt + $7\sqrt{(S/3) - 8}$. (Table 235-2 shows selected values up to 46 kV.

(2) For line conductors of No 2 AWG or larger:
clearance = 0.3 in per kilovolt + $8\sqrt{S/12}$. (Table 235-3 shows selected values up to 46 kV.)

(3) For voltages exceeding 814 kilovolts, the clearance shall be determined by the alternate method given by Rule 235B3.

Table 235-1. Minimum Horizontal Clearance at Supports Between Wires, Conductors, or Cables

(All voltages are between conductors involved except for railway feeders, which are to ground. The voltage between line conductors of different phases of different circuits shall be the vector difference of the voltages of both circuits. If the two conductors of different circuits are of like phase, the lower voltage conductor shall be considered grounded for the purpose of determining the clearance between them.)

Class of circuit	Clearance (in)	Notes
Open communication conductors	6	Preferable minimum. Does not apply at conductor transposition points.
	3	Permitted where pin spacings less than 6 in have been in regular use. Does not apply at conductor transposition points.
Railway feeders:		
0 to 750 V, No. 4/0 AWG or larger	6	Where 10 to 12 in clearance has already been established by practice, it may be continued, subject to the provisons of Rule 235B1b, for conductors having apparent sags not over 3 ft and for voltages not exceeding 8.7 kV.
0 to 750 V, smaller than No. 4/0 AWG	12	
750 V to 8.7 kV	12	
Supply conductors of the same circuit:		
0 to 8.7 kV	12	
8.7 to 50 kV	12 plus 0.4 per kV over 8.7 kV	
Above 50 kV	no value specified	
Supply conductors of different circuits:		
0 to 8.7 kV	12	For all voltages above 50 kV, the additional clearance shall be increased 3 percent for each 1000 ft in excess of 3300 ft above mean sea level. All clearances for voltages above 50 kV shall be based on the maximum operating voltage.
8.7 to 50 kV	12 plus 0.4 per kV over 8.7 kV	
50 to 814 kV	28.5 plus 0.4 per kV over 50 kV	

Table 235-2. Horizontal Clearances at Supports Between Line Conductors Smaller Than No. 2 AWG Based on Sags

Voltage between conductors (kV)	36	48	72	96	120	180	240	But not less than ①
			Horizontal clearance (in)					
2.4	14.7	20.5	28.7	35.0	40.3	51.2	60.1	12.0
4.16	15.3	21.1	29.3	35.6	40.9	51.8	60.7	12.0
12.47	17.7	23.5	31.7	38.0	43.3	54.2	63.1	13.5
13.2	18.0	23.8	32.0	38.3	43.6	54.5	63.4	13.8
13.8	18.1	23.9	32.1	38.4	43.7	54.6	63.5	14.0
14.4	18.3	24.1	32.3	38.6	43.9	54.8	63.7	14.3
24.94	21.5	27.3	35.5	41.8	47.1	58.0	66.9	18.5
34.5	24.4	30.2	38.4	44.7	50.0	60.9	69.8	22.4
46	27.8	33.6	41.8	48.1	53.4	64.3	73.2	26.9

The column heading above the data reads "Sag (in)".

① Clearance determined by Table 235-1, Rule 235B1a.

NOTE: Clearance = 0.3 in/kV + 7 $\sqrt{(S/3)} - 8$, where S is the sag in inches.

Table 235-3. Horizontal Clearances at Supports Between Line Conductors No. 2 AWG or Larger Based on Sags

Voltage between conductors (kV)	36	48	72	96	120	180	240	But not less than ①
			Horizontal clearance (in)					
2.4	14.6	16.7	20.2	23.3	26.0	31.7	36.5	12.0
4.16	15.1	17.3	20.8	23.8	26.5	32.2	37.0	12.0
12.47	17.6	19.7	23.6	26.3	29.0	34.7	39.5	13.5
13.2	17.8	20.0	23.5	26.5	29.2	34.9	39.7	13.8
13.8	18.0	20.1	23.7	26.7	29.4	35.1	39.9	14.0
14.4	18.2	20.3	23.8	26.9	29.6	35.3	40.1	14.3
24.94	21.3	23.5	27.0	30.0	32.8	38.4	43.2	18.5
34.5	24.2	26.4	29.9	32.9	35.6	41.3	46.1	22.4
46	27.7	29.8	33.3	36.4	39.1	44.8	49.6	26.9

The column heading above the data reads "Sag (in)".

① Clearance determined by Table 235-1, Rule 235B1a.

NOTE: Clearance = 0.3 in/kV + 8 $\sqrt{S/12}$, where S is the sag in inches.

(4) The clearance for voltages exceeding 50 kilovolts specified in Rule 235B1b(1) and (2) shall be increased 3% for each 1000 ft in excess of 3300 ft above mean sea level. All clearances for lines over 50 kilovolts shall be based on the maximum operating voltage.

2. Suspension Insulators

Where suspension insulators are used and are not restrained from movement, the clearance between conductors shall be increased so that one string of insulators may swing transversely throughout a range of insulator swing up to its maximum design swing angle without reducing the values given in Rule 235B1. The maximum design swing angle shall be based on a 6 pound per square foot wind on the conductor at final sag at $60°$F. This may be reduced to a 4 pound per square foot wind in areas sheltered by buildings, terrain, or other obstacles. The displacement of the wires, conductors, and cables shall include deflection of flexible structures and fittings, where such deflection would reduce the horizontal clearance between two wires, conductors, or cables.

3. Alternate Clearances for Different Circuits Where One or Both Circuits Exceed 98 Kilovolts, Alternating Current, to Ground or 139 Kilovolts Direct Current to Ground

The clearances specified in Rules 235B1 and 235B2 may be reduced for circuits with known switching surge factors but shall not be less than the clearances derived from the following computations.

a. Clearance

(1) The alternate basic clearance computed from the following equation and listed in Table 235-4 is the minimum electrical clearance between conductors of different circuits which shall be maintained under the expected loading conditions:

$$D = 3.28 \left[\frac{V_{L-L} \cdot (PU) \cdot a}{500 \, K} \right]^{1.667} b \quad \text{(feet)}$$

where

V_{L-L} maximum alternating current crest operating voltage in kilovolts between phases of different circuits or maximum

188

direct current operating voltage between poles of different circuits. If the phases are of the same phase and voltage magnitude one phase conductor shall be considered grounded;

PU maximum switching surge factor expressed in per-unit peak operating voltage between phases of different circuits and defined as a switching surge level between phases for circuit breakers corresponding to 98 percent probability that the maximum switching surge generated per breaker operation does not exceed this surge level, or the maximum anticipated switching surge level generated by other means, whichever is greater;

a = 1.15, the allowance for three standard deviations;

b = 1.03, the allowance for nonstandard atmospheric conditions;

K = 1.4, the configuration factor for a conductor-to-conductor gap.

(2) The value of *D* shall be increased 3 percent for each 1000 feet in excess of 1500 feet above mean sea level.

b. Limit

The clearance derived from Rule 235B3a shall not be less than the basic clearances given in Table 235-1 computed for 169 kilovolts alternating current.

C. Vertical Clearance Between Line Conductors

All line wires, conductors, and cables located at different levels on the same supporting structure shall have vertical clearances not less than the following.

1. Basic Clearance for conductors of same or different circuits.

The clearances given in Table 235-5 shall apply to line wires, conductors, or cables of 0 to 50 kV attached to supports. No value is specified for clearances between conductors of the same circuit exceeding 50 kV.

EXCEPTION 1: Line wires, conductors, or cables on vertical racks or separate brackets placed vertically and meeting the requirements of Rule 235G may have spacings as specified in that rule.

EXCEPTION 2: Where communication service drops cross under supply conductors on a common crossing structure, the clearance between the communication conductor and an effectively grounded supply conductor may be reduced to 4 in, provided the clearance between the communication conductor and supply conductors not effectively grounded meet the requirements of Rule 235C as appropriate.

Table 235-4. Electrical Clearances in Rule 235B3a(1)
(Add 3 percent for each 1000 ft in excess of 1500 ft above mean sea level.)

Maximum operating voltage phase to phase (kV)	Switching surge factor (per unit)	Switching surge (kV)	Electrical component of clearance (ft)
242	2.6 or less	890 or less	① 6.3
	2.8	958	7.2
	3.0	1027	8.1
	3.2 or more	1095 or more	② 8.8
362	1.8	893 or less	① 6.4
	2.0	1024	8.0
	2.2	1126	9.5
	2.4	1228	10.9
	2.6	1330	12.5
	2.7 or more	1382 or more	②12.8
550	1.6	1245	11.2
	1.8	1399	13.6
	2.0	1555	16.2
	2.2	1711	19.0
	2.3	1789 or more	②19.1
800	1.6	1810	20.8
	1.8	2037	25.3
	1.9 or more	2149 or more	②27.4

① Limited by Rule 235B3b.
② Need not be greater than specified in Rule 235B1 and 235B2.

190

EXCEPTION 3: Supply service drops of 0-750 V running above and parallel to communication service drops may have a minimum spacing of 12 inches at any point in the span including the point of and at their attachment to the building provided the non-grounded conductors are insulated and that a clearance of 40 inches is maintained between the two services at the pole.

EXCEPTION 4: This rule does not apply to conductors of the same circuit meeting Rule 230D.

2. Additional Clearances

Greater clearances than given in Table 235-5 (Rule 235C1) shall be provided under the following conditions. The increases are cumulative where more than one is applicable.

a. Voltages Exceeding 50 Kilovolts

(1) For voltages between 50 and 814 kilovolts, the clearance between conductors of different circuits shall be increased 0.4 in per kilovolt in excess of 50 kV.

EXCEPTION: For voltages to ground exceeding 98 kV alternating current or 139 kV direct current, clearances less than those required above are permitted for systems with known switching surge factors. (See Rule 235C3.)

(2) The increase in clearance for voltages in excess of 50 kV specified in Rule 235C2a(1) shall be increased 3% for each 1000 ft in excess of 3300 ft above mean sea level.

(3) All clearances for lines over 50 kV shall be based on the maximum operating voltage.

(4) No value is specified for clearances between conductors of the same circuit.

b. Conductors of Different Sags on Same Support

(1) Line conductors, supported at different levels on the same structure shall have vertical clearances at the supporting structures so adjusted that the minimum clearance at any point in the span shall be not less than any of the following with the upper conductor at its final unloaded sag at the maximum temperature for which the conductor is designed to operate and the lower conductor at its final unloaded sag at 60°F.

Table 235-5. Minimum Vertical Clearance at Supports Between Line Conductors

(All voltages are between conductors involved except for railway feeders, which are to ground.)

Conductors usually at lower levels	Open supply line conductors, 0 to 750 V; Supply cables of all voltages meeting Rule 230C1, 2, or 3; Neutral conductors meeting Rule 230E1 (in)	Open supply line conductors and cables of any voltage meeting Rule 230D②			
				15 to 50 kV	
		750 V to 8.7 kV (in)	8.7 to 15 kV (in)	Same utility (in)	Different utilities (in)
Communication conductors:					
General	①40	③40	60	40	60
Used in operation of supply lines	16	16	40	40	60
Supply conductors:					
0 to 750 V; Supply cables meeting Rule 230C1, 2, or 3; Neutral conductors meeting Rule 230E1	16	④16	40	40	60
750 V to 8.7 kV		④16	40	40	60
8.7 kV to 15 kV: If worked on alive with live line tools, and adjacent circuits are neither de-energized nor covered with shields or protectors.			40	40	60

192

If not worked on alive except when adjacent circuits (either above or below) are de-energized or covered by shields or protectors, or by the use of live line tools not requiring linemen to go between live wires

Exceeding 15 kV, but not exceeding 50 kV

16	⑤40	⑤40
	⑤40	⑤40

① Where supply circuits of 600 V or less, with transmitted power of 5000 W or less, are run below communication circuits in accordance with Rule 220B2 the clearance may be reduced to 16 in.

② A conductor which is effectively grounded throughout its length, and is assocated with a supply circuit of 0 to 22 kV may have the clearances specified for cables meeting Rule 230C1, 2 or 3.

③ This shall be increased to 40 in when the communication conductors are carried above supply conductors unless the communication-line-conductor size is that required for grade C supply lines.

④ Where conductors are operated by different utilities, a minimum vertical clearance of 40 in is recommended.

⑤ These values do not apply to conductors of the same circuit or circuits being carried on adjacent conductor supports.

(a) For voltages less than 50 kilovolts between conductors, 75 percent of that required at the supports by Table 235-5.

(b) For voltages more than 50 kilovolts between conductors, the value specified in Rule 235C2b(1)(a) increased in accordance with Rule 235C2a.

(2) Sags should be readjusted when necessary to accomplish the foregoing, but not reduced sufficiently to conflict with the requirements of Rule 261H2. In cases where conductors of different sizes are strung to the same sag for the sake of appearance or to maintain unreduced clearance throughout storms, the chosen sag should be such as will keep the smallest conductor involved in compliance with the sag requirements of Rule 261H2.

(3) For span lengths in excess of 150 feet, vertical clearance at the structure between open supply conductors and communication cables or contors shall be adjusted so that under conditions of conductor temperature of 60°F, no wind and final unloaded sag, no supply conductor of 750 volts or less shall be lower in the span than a straight line joining the points of support of the highest communication cable or conductor, and no supply conductor of over 750 volts but less than 50 kilovolts shall be lower in the span than 30 inches above such a straight line.

EXCEPTION: Effectively grounded supply conductors associated with systems of 50 kilovolts or less need only meet the provisions of Rule 235C2b(1).

3. **Alternate Clearances for Different Circuits Where One or Both Exceed 98 kilovolts, Alternating Current, or 139 kilovolts Direct Current to Ground**
The clearances specified in Rules 235C1 and 235C2 may be reduced for circuits with known switching surge factors, but shall not be less than the crossing clearances required by Rule 233C3.

D. Diagonal Clearance Between Line Wires, Conductors, and Cables Located at Different Levels on the Same Supporting Structure

No wire, conductor, or cable may be closer to any other wire, conductor, or cable than defined by the dashed line in Fig 235-1, where V and H are determined in accordance with other parts of Rule 235.

E. Clearances in Any Direction from Line Conductors to Supports, and to Vertical or Lateral Conductors, Span or Guy Wires Attached to the Same Support

1. Fixed Supports
   Clearances shall not be less than given in Table 235-6.

   EXCEPTION: For voltages exceeding 98 kilovolts alternating current to ground or 139 kilovolts direct current to ground, clearances less than those required by Table 235-6 are permitted to systems with known switching surge factor. (See Rule 235E3.)

2. Suspension Insulators
   Where suspension insulators are used and are not restrained from movement, the clearance shall be in-

Fig 235-1
Clearance Diagram for Energized Conductor

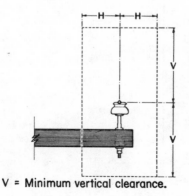

V = Minimum vertical clearance.

H = Minimum horizontal clearance.

Table 235-6. Minimum Clearance in Any Direction from Line Conductors to Supports and to Vertical or Lateral Conductors, Span, or Guy Wires Attached to the Same Support

	Communication lines		Supply lines		
			Circuit phase-to-phase voltage		
Clearance of line conductors from	In general (in)	On jointly used structures (in)	0 to 8.7 kV (in)	8.7 to 50 kV (in)	50 to 814 kV④⑨ (in)
Vertical and lateral conductors:					
Of the same circuit	3	3	3	3 plus 0.25 per kV over 8.7 kV	no value specified
Of other circuits	3	3	⑤6	6 plus 0.4 per kV over 8.7 kV	23 plus 0.4 per kV over 50 kV
Span or guy wires, or messengers attached to same structure:					
When parallel to line	⑦3	①⑦6	①12	12 plus 0.4 per kV over 8.7 kV	29 plus 0.4 per kV over 50 kV
Anchor guys	⑦3	①⑦6	①6	6 plus 0.25 per kV over 8.7 kV	16 plus 0.25 per kV over 50 kV
All other	⑦3	①⑦6	6	6 plus 0.4 per kV over 8.7 kV	23 plus 0.4 per kV over 50 kV

Surface of support arms	②3	②3	⑥⑧3	3 plus 0.2 per kV over 8.7 kV⑥⑧⑩	11 plus 0.2 per kV over 50 kV
Surface of structures:					
On jointly used structures	—	②5	③⑥⑧5	5 plus 0.2 per kV over 8.7 kV⑥⑧	13 plus 0.2 per kV over 50 kV
All other	②3	—	⑥⑧3	3 plus 0.2 per kV over 8.7 kV⑥⑧	11 plus 0.2 per kV over 50 kV

①For guy wires, if practical. For clearances between span wires and communication conductors, see Rule 238C.

On jointly used structures, guys which pass within 12 in of supply conductors, and also pass within 12 in of communication cables, shall be protected w'th a suitable insulating covering where the guy passes the supply conductors, unless the guy is effectively grounded or insulated with a strain insulator at a point below the lowest supply conductor and above the highest communication cable.

② Communication conductors may be attached to supports on the sides or bottom of crossarms or surfaces of poles with less clearances.

③This clearance applies only to supply conductors at the support below communication conductors, on jointly used structures.

Where supply conductors are above communication conductors, this clearance may be reduced to 3 in except for supply conductors of 0 to 750 V whose clearance may be reduced to 1 in.

④All clearances for line over 50 kV shall be based on the maximum operating voltage. For voltages exceeding 814 kV, the clearance shall be determined by the alternate method given by Rule 235E3.

⑤For supply circuits of 0 to 750 V, this clearance may be reduced to 3 in.

⑥A neutral conductor meeting Rule 230E1 may be attached directly to the structure surface.

⑦Guys and messengers may be attached to the same strain plates or to the same through bolts.

⑧For open supply circuits of 0 to 750 V and supply cables of all voltages meeting Rule 230C1, 2 or 3, this clearance may be reduced to 1 in.

⑨The additional clearance for voltages in excess of 50 kV specified in Table 235-6 shall be increased 3 percent for each 1000 ft in excess of 3300 ft above mean sea level.

⑩ Where circuit is effectively grounded and neutral conductor meets Rule 230E1, phase-to-neutral voltage shall be used to determine clearance from phase conductor to surface of support arms.

creased so that the string of insulators may swing
transversely throughout a range of insulator swing up to
its maximum design swing angle without reducing the
values given in Rule 235E1. The maximum design swing
angle shall be based on a 6 pound per square foot wind
on the conductor at final sag at 60°F. This may be
reduced to a 4 pound per square foot wind in areas
sheltered by buildings, terrain, or other obstacles. The
displacement of the wires, conductors, and cables shall
include deflection of flexible structures and fittings,
where such deflection would reduce the clearance.

3. Alternate Clearances for Voltages Exceeding 98 kV
   Alternating Current to Ground or 139 kV Direct Current
   to Ground
   The clearances specified in Rules 235E1 and 235E2 may
   be reduced for circuits with known switching surge
   factors but shall not be less than the following.

   a. Alternate Clearances to Anchor Guys, and Vertical
      or Lateral Conductors
      The alternate clearances shall not be less than the
      crossing clearances required by Rule 233A3 for the
      conductor voltages concerned. For the purpose of
      this rule, anchor guys shall be assumed to be at
      ground potential.

   b. Alternate Clearance to Lightning Protection Wires
      The alternate clearances shall not be less than the
      crossing clearances required by Rule 233A3 for the
      conductor voltage concerned. For the purpose of
      this rule, lightning protection wires shall be assumed
      to be at ground potential.

   c. Alternate Clearance to Surface of Support Arms and
      Structures
      (1) Alternate Clearance
          (a) Basic Computation
              The alternate clearance computed from the
              following equation is the minimum electrical
              clearance which shall be maintained under
              the expected loading conditions. For con-
              venience, clearances for typical system volt-
              ages are shown in Table 235-7.

$$D = 39.37 \left[\frac{V \cdot (PU) \cdot a}{500\,K} \right]^{1.667} b \qquad \text{(inches)}$$

where

V maximum alternating current crest operating voltage to ground or maximum direct current operating voltage to ground in kilovolts;

PU maximum switching surge factor expressed in per-unit peak voltage to ground and defined as a switching surge level for circuit breakers corresponding to 98 percent probability that the maximum switching surge generated per breaker operation does not exceed this surge level, or the maximum anticipated switching surge level generated by other means, whichever is greater;

a $\Bigg\{$ = 1.15, the allowance for three standard deviations with fixed insulator supports;

         = 1.05, the allowance for one standard deviation with free swinging insulators;

b = 1.03, the allowance for nonstandard atmospheric conditions;

K = 1.2, the configuration factor for conductor-to-tower window.

(b) Atmospheric Correction

The value of D shall be increased 3 percent for each 1000 feet in excess of 1500 feet above mean sea level.

(2) Limits

The alternate clearance shall not be less than the clearance of Table 235-6 for 169 kV alternating current. The alternate clearance shall be checked for adequacy of clearance to workmen and increased, if necessary, where work is to be done on the structure while the circuit is energized. (Also see NESC, Part 4.)

F. **Clearance Between Supply Circuits of Different Voltage Classifications on the Same Support Arm**

Supply circuits of any one voltage classification as given in Table 235-5 may be maintained on the same support arm with supply circuits of the next consecutive voltage classification only under one or more of the following conditions:

1. If they occupy positions on opposite sides of the structure.
2. If in bridge-arm or sidearm construction, the clearance is not less than the climbing space required for the higher voltage concerned and provided for in Rule 236.

Table 235-7. Minimum Clearance in Any Direction from Line Conductors to Supports

Maximum operating voltage phase to phase (kV)	Switching surge factor (per unit)	Switching surge (kV)	Minimum clearance to supports	
			Fixed (in)	Free swinging at maximum angle (in)
242	2.4	474	① 35	① 35
	2.6	514	40	① 35
	2.8	553	45	38
	3.0	593	② 50	43
	3.2	632	② 50	48
362	1.6	473	① 35	① 35
	1.8	532	42	36
	2.0	591	50	48
	2.2	650	59	51
	2.4	709	68	59
	2.5	739	② 73	63
550	1.6	719	70	60
	1.8	808	85	73
	2.0	898	101	87
	2.2	988	② 111	101
800	1.6	1045	130	111
	1.8	1176	158	135
	1.9	1241	② 161	148
	2.0	1306	② 161	② 161

① Limited by Rule 235E3(c)(2).
② Need not be greater than specified in Rules 235E1 and 2.

3. If the higher voltage conductors occupy the outer positions and the lower voltage conductors occupy the inner positions.

4. If series lighting or similar supply circuits are ordinarily dead during periods of work on or above the support arm concerned.

5. If the two circuits concerned are communication circuits used in the operation of supply lines, and supply circuits of less than 8.7 kilovolts, and are owned by the same utility, provided they are installed as specified in Rule 235F1 or 235F2.

G. Conductor Spacing: Vertical Racks

Conductors or cables may be carried on vertical racks or separate brackets other than wood placed vertically on one side of the structure and securely attached thereto with less clearance between the wires, conductors, or cables than specified in Rule 235C if all the following conditions are met:

1. The voltage shall not be more than 750 volts, except supply cables and conductors meeting Rules 230C1 or 230C2 which may carry any voltage.

2. Conductors shall be of the same material or materials, except that different materials may be used if their sag-tension characteristics and arrangement are such that the spacing specified in Rule 235C3 is maintained under all service conditions.

3. Vertical spacing between conductors shall be not less than the following:

Span length (feet)	Vertical spacing between conductors (inches)
0 to 150	4
150 to 200	6
200 to 250	8
250 to 300	12

EXCEPTION: The vertical spacing may be reduced where the conductors are held apart by intermediate spacers, but may not be less than 4 inches.

236. Climbing Space

The following requirements apply only to portions of structures which workmen ascend.

A. Location and Dimensions

1. A climbing space having the horizontal dimensions specified in Rule 236E shall be provided past any conductors, support arms, or other parts.

2. The climbing space need be provided on one side or corner of the support only.

3. The climbing space shall extend vertically past any conductor or other part between levels above and below the conductor as specified in Rules 236E, F, G, and I, but may otherwise be shifted from any side or corner of the support to any other side or corner.

B. Portions of Supporting Structures in Climbing Space

Portions of the supporting structure when included in one side or corner of the climbing space are not considered to obstruct the climbing space.

C. Support Arm Location Relative to Climbing Space

RECOMMENDATION: Support arms should be located on the same side of the pole.

EXCEPTION: This recommendation does not apply where double crossarms are used on any pole or where crossarms on any pole are not all parallel.

D. Location of Supply Apparatus Relative to Climbing Space

Supply apparatus including, but not limited to, transformers, regulators, capacitors, cable terminals (potheads), surge arresters, and switches when located below conductors or other attachments shall be mounted outside of the climbing space.

E. Climbing Space Between Conductors

Climbing space between conductors shall be of the horizontal dimensions specified in Table 236-1. These dimensions are intended to provide a clear climbing space of 24 inches while the conductors bounding the climbing space are covered with temporarily installed protective covering rated for the voltage involved. The climbing space shall be provided both along and across the line, and shall be projected vertically not less than 40 inches above and below the limiting conductors. Where communication conductors are above supply conductors of more than 8.7 kilovolts to ground or 15 kilovolts line to line, the climbing space shall be projected vertically at least 60 inches above the highest supply conductors.

EXCEPTION 1: This rule does not apply if it is the unvarying practice of the employers concerned to prohibit employees from ascending beyond the conductors or equipment of a given line or structure unless the conductors or equipment are de-energized.

EXCEPTION 2: For supply conductors carried on a structure in a position below communication facilities in the manner permitted in Rule 220B2, the climbing space need not extend more than 2 feet above such supply space.

EXCEPTION 3: If the conductors are owned, operated, or maintained by the same utility, the climbing space may be provided by temporarily moving the line conductors using live line tools.

F. **Climbing Space on Buckarm Construction**
The full width of climbing space shall be maintained on buckarm construction and shall extend vertically in the same position at least 40 inches (or 60 inches where required by Rule 236E) above and below any limiting conductor.

Method of Providing Climbing Space on Buckarm Construction

With circuits of less than 8.7 kilovolts to ground or 15 kilovolts line to line and span lengths not exceeding 150 feet and sags not exceeding 15 inches for wires of No. 2 AWG and larger sizes, or 30 inches for wires smaller than No. 2 AWG, a six-pin crossarm having pin spacing of 14½ inches may be used to provide a 30 inch climbing space on one corner of a junction pole by omitting the pole pins on all arms, and inserting pins midway between the remaining pins so as to give a spacing of 7¼ inches, provided that each conductor on the end of every arm is tied to the same side of its insulator, and that the spacing on the next pole is not less than 14½ inches.

G. **Climbing Space Past Longitudinal Runs Not on Support Arms**
The full width of climbing space shall be provided past longitudinal runs and shall extend vertically in the same position from 40 inches below the run to a point 40 inches above (or 60 inches where required by Rule 236E). The width of climbing space shall be measured from the longitudinal run concerned. Longitudinal runs on racks, or cables on messengers, are not considered as obstructing the climbing space if all wires concerned are covered by rubber protective equipment or otherwise guarded as an unvarying practice before

Table 236-1. Minimum Horizontal Clearance Between Conductors Bounding the Climbing Space

(All voltages are between the two conductors bounding the climbing space except for communications conductors which are voltage to ground. Where the two conductors are in different circuits, the voltage between conductors shall be the arithmetic sum of the voltages of each conductor to ground for a grounded circuit or phase to phase for an ungrounded circuit.)

| Character of conductors adjacent to climbing space | Voltage of conductors | Horizontal clearance between conductors bounding the climbing space ③ | | | |
| | | On structures used solely by | | On jointly used structures | |
		Communication conductors (in)	Supply conductors (in)	Supply conductors above communication conductors (in)	Communication conductors above supply conductors ① (in)
Communication conductors	0 to 150 V	no requirements	—	②	no requirements
	exceeding 150 V	24 recommended	—	②	24 recommended
Supply cables meeting Rule 230C1	all voltages	—	—	②	no requirement
Supply cables meeting Rule 230C2 or 3	all voltages	—	24	24	30
Open supply line conductors and supply cables meeting Rule	0 to 750 V	—	24	24	30
	750 V to 15 kV	—	30	30	30
	15 kV to 28 kV	—	36	36	36
	28 kV to 38 kV	—	40	40	

204

230D			
38 kV to 50 kV	—	46	46
50 kV to 73 kV	—	54	54
exceeding 73 kV	—	more than 54	

① This relation of levels is not, in general, desirable and should be avoided.

② Climbing space shall be the same as required for the supply conductors immediately above, with a maximum of 30 in, except that a climbing space of 16 in across the line may be employed for communication cables or conductors where the only supply conductors at a higher level are secondaries (0 to 750 V) supplying airport or airway marker lights or crossing over the communication line and attached to the pole top or to a pole top extension fixture.

③ Attention is called to the operating requirements of Rule 422B and 427C, Part 4, of this code.

workmen climb past them. This does not apply where communication conductors are above the longitudinal runs concerned.

EXCEPTION 1: If a supply longitudinal run is placed on the side or corner of the supporting structure where climbing space is provided, the width of climbing space shall be measured horizontally from the center of the structure to the nearest supply conductors on support arms, under both of the following conditions:

(i) Where the longitudinal run consists of open supply conductors carrying not more than 750 volts, or supply cables and conductors meeting Rule 230C, all voltages; and is supported close to the structure as by brackets, racks, or pins close to the structure.

(ii) Where the nearest supply conductors on support arms are parallel to and on the same side of the structure as the longitudinal run and within 4 feet above or below the run.

EXCEPTION 2: For supply conductors carried on a structure in a position below communication facilities in the manner permitted in Rule 220B2, the climbing space need not extend more than 2 feet above such supply space.

H. Climbing Space Past Vertical Conductors
Vertical runs physically protected by suitable conduit or other protective covering and securely attached without spacers to the surface of the line structure are not considered to obstruct the climbing space.

I. Climbing Space Near Ridge-Pin Conductors
The climbing space specified in Table 236-1 shall be provided above the top support arm to the ridge-pin conductor but need not be carried past it.

237. Working Space

A. Location of Working Spaces
Working spaces shall be provided on the climbing face of the structure at each side of the climbing space.

B. Dimensions of Working Spaces
1. Along the Support Arm
The working space shall extend from the climbing space to the outmost conductor position on the support arm.
2. At Right Angles to the Support Arm
The working space shall have the same dimension as the climbing space (see Rule 236E). This dimension shall be measured horizontally from the face of the support arm.

3. Vertically
The working space shall have a height not less than that required by Rule 235 for the vertical separation of line conductors carried at different levels on the same support.

C. Location of Vertical and Lateral Conductors Relative to Working Spaces
The working spaces shall not be obstructed by vertical or lateral conductors. Such conductors shall be located on the opposite side of the pole from the climbing side or on the climbing side of the pole at a distance from the support arm at least as great as the width of climbing space required for the highest voltage conductors concerned. Vertical conductors enclosed in suitable conduit may be attached on the climbing side of the structure.

D. Location of Buckarms Relative to Working Spaces
Buckarms may be used under any of the following conditions, provided the climbing space is maintained. Climbing space may be obtained as in Rule 236F.

1. Standard Height of Working Space
Lateral working space of the height required by Table 235-5 shall be provided between the lateral conductors attached to the buckarm and the line conductors. This may be accomplished by increasing the spacing between the line support arms as shown in Figure 237-1.

Fig 237-1
Obstruction of Working Space by Buckarm

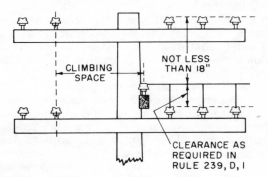

2. Reduced Height of Working Space

Where no circuits exceeding 8.7 kilovolts to ground or 15 kilovolts line to line are involved, and the clearances of Rules 235B1a and 235B1b are maintained, conductors supported on buckarms may be placed between line conductors having normal vertical spacing, even though such buckarms obstruct the normal working space, provided that a working space of not less than 18 inches in height is maintained either above or below line conductors and buckarm conductors.

EXCEPTION: The above working space may be reduced to 12 inches if both of the following conditions exist:

(i) Not more than two sets of line arms and buckarms are involved.

(ii) Working conditions are rendered safe by providing rubber protective equipment or other suitable devices to insulate and cover line conductors and equipment which are not being worked upon.

238. Vertical Clearance Between Line Wires, Conductors, or Cables and Noncurrent-Carrying Metal Parts of Equipment Located at Different Levels on the Same Structure

A. Equipment

For the purpose of measuring clearances under this rule, "equipment" shall be taken to mean noncurrent-carrying metal parts of equipment, including metal supports for cables or conductors, and metal support braces which are attached to metal supports or are less than 1 inch from transformer cases or hangers which are not effectively grounded.

B. Clearances in General

Vertical clearances between supply conductors and communication equipment, between communication conductors and supply equipment, and between supply and communication equipment shall be as specified in Table 238-1 except as provided in Rule 238C.

C. Clearances for Span Wires or Brackets

Span wires or brackets carrying luminaires or trolley conductors shall have at least the vertical clearances in inches from communication equipment set forth in Table 238-2.

D. Clearance from Drip Loops of Luminaire Brackets

If a drip loop of conductors entering a luminaire bracket from the surface of the structure is above a communication cable, the lowest point of the loop shall be at least 12 inches above communication cable or through bolt.

Table 238-1. Vertical Clearance Between Conductors and
Noncurrent Carrying Metal Parts of Equipment
(Voltages are phase to ground for effectively grounded circuits
and those other circuits where all ground faults are cleared by
promptly de-energizing the faulted section, both initially and
following subsequent breaker operations. See the definition
section for voltages of other systems.)

Supply voltage (kV)	Vertical clearance (in)
0 to 8.7	① 40
8.7 to 50	① 60
over 50	60 plus 0.4 per KV over 50 kV

①Where noncurrent carrying parts of equipment are effectively
grounded consistently throughout well-defined areas and where communication is at lower levels, clearances may be reduced to 30 in.

239. Clearances of Vertical and Lateral Conductors from Other Wires and Surfaces on the Same Support

Vertical and lateral conductors shall have the clearances and
separations required by this rule from other conductors,
wires, or surfaces on the same support.

EXCEPTION 1: This rule does not prohibit the placing of
supply circuits of the same or next voltage classification in
the same duct, if each circuit or set of wires is enclosed in
a metal sheath.

EXCEPTION 2: This rule does not prohibit the placing of
paired communication conductors in rings attached directly
to the structure or to messenger.

EXCEPTION 3: This rule does not prohibit placing grounding conductors, neutral conductors meeting Rule 230E1,
supply cables meeting Rule 230C1, or conductors physically protected by enclosing in conduit, directly on the support.

EXCEPTION 4: This rule does not prohibit placing properly insulated supply circuits of 600 volts or less and not exceeding 5000 watts in the same cable with control circuits
with which they are associated.

A. Location of Vertical or Lateral Conductors Relative to Climbing Spaces, Working Spaces, and Pole Steps

Vertical or lateral conductors shall be located so that they do
not obstruct climbing spaces, or lateral working spaces between line conductors at different levels, or interfere with
the safe use of pole steps.

Table 238-2.   Vertical Clearance of Span Wires and
Brackets from Communications Lines

	Carrying luminaires		Carrying trolley conductors	
	Not effectively grounded (in)	Effectively grounded (in)	Not effectively grounded (in)	Effectively grounded (in)
Above communication support arms	①20	①20	①20	①20
Below communication support arms	③40	24	24	24
Above messengers carrying communication cables	①20	4	12	4
Below messengers carrying communication cables	④40	4	12	4
From terminal box of communication cable	①20	4	②12	4
From communication brackets, bridle wire rings, or drive hooks	①16	4	4	4

① This may be reduced to 12 in for either span wires or metal parts of brackets at points 40 in or more from the structure surface.

② Where it is not practical to obtain a clearance of 1 ft from terminal boxes of communication cables, all metal parts of terminals shall have the greatest possible separation from fixtures or span wires including all supporting screws and bolts of both attachments.

③ This may be reduced to 24 in for luminaires operating at less than 150 V to ground.

④ This may be reduced to 20 inches for luminaires operating at less than 150 volts to ground.

EXCEPTION 1: This rule does not apply to portions of the structure which workmen do not ascend while the conductors in question are alive.

B. Conductors Not in Conduit

Conductors not encased in conduit shall have the same clearances from conduits as from other surfaces of structures.

C. Mechanical Protection Near Ground

Where within 8 feet of the ground, all vertical conductors, cables, and grounding wires shall be protected by a covering which gives suitable mechanical protection. For grounding wires from surge arresters, the protective covering just specified shall be of wood molding or of other nonmetallic material giving equivalent mechanical protection.

EXCEPTION 1: This covering may be omitted from armored cables or cables installed in a grounded metal conduit.

EXCEPTION 2: This covering may be omitted from lead-sheathed cables used in rural districts.

EXCEPTION 3: This covering may be omitted from vertical runs of communication cables or conductors.

EXCEPTION 4: This covering may be omitted from grounding wires used in rural districts or in any area where the grounding wire is one of a number of grounding wires used to provide multiple grounds.

EXCEPTION 5: This covering may be omitted from wires which are used solely to protect poles from lightning.

D. Requirements for Vertical and Lateral Supply Conductors on Supply Line Structures or Within Supply Space on Jointly Used Structures

1. General Clearances

   In general, clearances shall be not less than the values specified in Table 239-1 or Rule 235E.

2. Special Cases

   The following requirements apply only to portions of a structure which workmen ascend while the conductors in question are alive.

   a. Sidearm Construction

      Vertical conductors in cables meeting Rule 230C1 and grounding wires may be run without insulating protection from supply line conductors on structures used only for supply lines and employing sidearm construction on the side of the structure opposite to the line conductors if climbing space is provided on the line-conductor side of the structure.

Table 239-1.   Clearance of Vertical and Lateral Conductors
(Circuit Phase-to-Phase Voltage)

Clearance of vertical and lateral conductors	0 to 8.7 kV (in)	8.7 to 50 kV (in)	Over 50 kV④ (in)
From surfaces of supports	①②3	3 plus 0.2 per kV over 8.7 kV	11 plus 0.2 per kV over 50 kV
From span, guy, and messenger wires	6	6 plus 0.4 per kV over 8.7 kV③	23 plus 0.4 per kV over 50 kV③

① A neutral conductor meeting Rule 230E1 may be attached directly to the structure surface.

② For supply circuits of 0 to 750 V this clearance may be reduced to 1 in.

③ Multiplier may be reduced to 0.25 in/kV for anchor guys.

④ The additional clearance for voltages in excess of 50 kV specified in Table 239-1 shall be increased 3 percent for each 1000 ft in excess of 3300 ft above mean sea level.

Table 239-2.   Clearances Between Open Vertical Conductors and Pole Center

(Voltages are phase to ground for effectively grounded circuits and those other circuits where all ground faults are cleared by promptly de-energizing the faulted section, both initially and following subsequent breaker operations. See the definition section for voltages of other systems.)

Voltage (kV)	Distance above and below open supply conductors where clearances apply (ft)	Minimum clearance between vertical conductor and pole center (in)
0 to 8.7	4	15
8.7 to 16	6	20
16 to 22	6	23
22 to 30	6	26
30 to 50	6	34

b. Conductors to Luminaires

On structures used only for supply lines, open wires may be run from the supply line arm directly to the head of a luminaire, provided the clearances of Table 239-1 are obtained and the open wires are substantially supported at both ends.

c. Conductors of Less Than 300 Volts

Vertical or lateral secondary supply conductors of not more than 300 volts to ground may be run in multiple-conductor cable attached directly to the structure surface or to support arms in such a manner as to avoid abrasion at the point of attachment. Each conductor of such cable which is not effectively grounded, or the entire cable assembly, shall have an insulating covering required for a conductor of at least 600 volts.

d. Other Conditions

If open wire conductors are within 4 feet of the pole, vertical conductors shall be run in one of the following ways.

(1) Open vertical conductors shall have the clearances given in Table 239-2 within the zone specified in the table.

(2) Within the zone above and below open supply conductor as given in Table 239-2 vertical and lateral conductors may be enclosed in nonmetallic conduit, or in cable protected by an insulating covering and may be run on the pole surface.

(3) Grounding conductors may be run on the pole surface without molding.

E. Requirements for Vertical and Lateral Communications Conductors on Communication Line Structures or Within the Communication Space on Jointly Used Structures

1. Clearances from Wires

The clearances of uninsulated vertical and lateral conductors from other conductors (except those in the same ring run) and from guy, span, or messenger wires shall be 3 inches.

2. Clearances from Supporting Structure Surfaces

Vertical and lateral insulated communication conductors may be attached directly to a structure. They shall have a vertical clearance of at least 40 inches from any supply conductors (other than vertical runs or luminaire leads) of 8.7 kilovolts or less, or 60 inches if more than 8.7 kilovolts.

EXCEPTION: These clearances do not apply where the supply circuits involved are those carried in the manner specified in Rule 220B2.

F. Requirements for Vertical Supply Conductors Passing Through Communication Space on Jointly Used Line Structures

1. Grounded Metal-Sheathed Cables
Grounded metal-sheathed cables may be fastened directly to the surface of the line structure. Such cables shall be protected with suitable nonmetallic covering when the line structure also carries trolley attachments or when an ungrounded luminaire is attached below the communication cable. The grounded metal-sheathed cable shall be protected with a non metallic covering for a distance of 40 inches above the highest communication wire and 6 feet below the lowest trolley attachment or ungrounded luminaire fixture.

2. Jacketed Multiple-Conductor Cables
Jacketed multiple-conductor cables operating at voltages not exceeding 300 volts to ground may be attached directly to the surface of the line structure. Each conductor shall be insulated for a potential of at least 600 volts. Where used as aerial services, the point where such cables leave the structure shall be at least 40 inches above the highest or 40 inches below the lowest communication attachment. All splices and connections in the cable shall be insulated. No additional protection is required.

3. Grounded Metal Covering
Conductors of all voltages may be run in effectively grounded metal covering. Such metal covering shall be protected with a nonmetallic covering under the same conditions and to the same extent as required for grounded metal-sheathed cables in Rule 239F1.

4. Suspended from Supply Support Arm
Lamp leads of lighting circuits may be run from supply support arms directly to a bracket or luminaire under the following conditions:
a. The vertical run shall consist of paired wires or multiple-conductor cable securely attached at both ends to suitable brackets and insulators.

    b. The vertical run shall be held taut at least 40 inches from the surface of the pole through the communication space at least 12 inches beyond the end of any communication support arm by which is passes, and at least 6 inches from communication drop wires, and at least 20 inches from any communication cable.

    c. Insulators attached to luminaire brackets for supporting the vertical run shall be capable of meeting, in the position in which they are installed, the same flashover requirements as the luminaire insulators.

    d. Each conductor of the vertical run shall be No. 10 AWG or larger.

5. Supply Grounding Conductors

    a. Supply grounding conductors may be run bare where there are no trolley attachments or ungrounded street lighting fixtures, or both, located below the communication attachment provided:

      (1) the grounding conductor is directly (metallically) connected to a conductor which forms part of an effective grounding system, and

      (2) the grounding conductor has no connection to supply equipment between the grounding electrode and the effectively grounded conductor unless the supply equipment has additional connections to the effectively grounded conductor.

    b. Supply grounding conductors not conforming to Rule 239F5a shall be protected with a suitable nonmetallic covering to the same extent as required for grounded metal-sheathed cables in Rule 239F1.

6. Clearance from Through Bolts

Vertical runs of supply conductors or cables shall have a clearance of not less than 2 inches from exposed through bolts and other exposed metal objects attached thereto which are associated with communication line equipment.

EXCEPTION: Vertical runs of effectively grounded supply conductors may have a clearance of 1 inch from the end of exposed communication through bolts.

G. Requirements for Vertical Communication Conductors Passing Through Supply Space on Jointly Used Structures.

All vertical runs of communication conductors passing through supply space shall be installed as follows.

1. **Metal-Sheathed Communication Cables**
    Vertical runs of metal-sheath communication cables shall be covered with wood molding, or other suitable non-metallic material, where they pass trolley feeders or other supply line conductors. This nonmetallic covering shall extend from a point 40 inches above the highest trolley feeders, or other supply conductors, to a point 6 feet below the lowest trolley feeders or other supply conductors, but need not extend below the top of any mechanical protection which may be provided near the ground.

    EXCEPTION: Communication cables may be run vertically on the pole through space occupied by railroad signal supply circuits in the lower position, as permitted in Rule 220B2, without nonmetallic covering within the supply space.

2. **Communication Conductors**
    Vertical runs of insulated communication conductors shall be covered with wood molding, or other suitable nonmetallic material, to the extent required for metal-sheathed communication cables in Rule 239G1, where such conductors pass trolley feeders or other supply conductors.

    EXCEPTION: Communication conductors may be run vertically on the structure through space occupied by railroad-signal supply circuits in the lower position, as permitted in Rule 220B2, without nonmetallic covering within the supply space.

3. **Communication Grounding Conductors**
    Vertical communication grounding conductors shall be covered with wood molding or other nonmetallic material between points at least 6 feet below and 40 inches above any trolley feeders or other supply line conductors by which they pass.

    EXCEPTION: Communication grounding conductors may be run vertically on the structure through space occupied by railroad-signal supply circuits in the lower position, as permitted in Rule 220B2, without nonmetallic covering within the supply space.

4. **Separation from Through Bolts**
    Vertical runs of communication conductors shall have a clearance of one-eighth of the pole circumference but

not less than 2 inches from through bolts and other metal objects attached thereto which are associated with supply line equipment.

EXCEPTION: Vertical runs of effectively grounded communications conductors may have a separation of 1 inch from the end of supply through bolts.

Section 24. Grades of Construction

240. General

A. The grades of construction are specified in this section on the basis of the required strengths for safety. Where two or more conditions define the grade of construction required, the grade used shall be the highest one required by any of the conditions.

B. For the purposes of this section, the voltage values for direct-current circuits shall be considered equivalent to the rms values for alternating-current circuits.

241. Application of Grades of Construction to Different Situations

A. Supply Cables

For the purposes of these rules, supply cables are classified by two types as follows:

Type 1

Supply cables conforming to Rules 230C1, 230C2, or 230C3 shall be installed in accordance with Rule 261I1.

Type 2

All other supply cables are required to have the same grade of construction as open-wire supply conductors of the same voltage.

B. Order of Grades

The relative order of grades for supply and communication conductors and supporting structures is B, C, and N, grade B being the highest. Grade D is specified only for communication lines, and here it is higher than Grade N. Grade D cannot be directly compared with Grades B and C, but Rule 241C3b provides for conditions when such a combination of construction requirements exists.

217

C. **At Crossings**
1. Grade of Upper Line
Conductors and supporting structures of a line crossing over another line shall have the grade of construction specified in Rules 241C3, 242, and 243.
2. Grade of Lower Line
Conductors and supporting structures of a line crossing under another line need only have the grades of construction which would be required if the line at the higher level were not there.
3. Multiple Crossings
a. Where a line crosses in one span over two or more other lines, or where one line crosses over a span of a second line, which span in turn crosses a span of a third line, the grade of construction of the uppermost line shall be not less than the highest grade which would be required of either one of the lower lines when crossing the other lower line.
b. Where communication conductors cross over supply conductors and railroad tracks in the same span, the grades of construction shall be in accordance with those listed in Table 241-1. It is recommended that the placing of communication conductors above supply conductors generally be avoided unless the supply conductors are trolley-contact conductors and their associated feeders.

Table 241-1.   Grades of Construction for Communication Conductors Crossing Over Railroad Tracks and Supply Lines

When crossing over	Communication conductor grades
Railroad tracks and supply lines of 0 to 750 V to ground, or Type 1 supply cables of all voltages	D
Railroad tracks and supply lines exceeding 750 V to ground	B

D. **Conflicts** *(see definitions)*

The grade of construction of the conflicting structure shall be as required by Rule 243A5.

242. **Grades of Construction for Conductors**

The grades of construction required for conductors are given in Tables 242-1 and 242-2. For the purpose of these tables certain classes of circuits are treated as follows:

A. **Constant-Current Circuit Conductors**

The grade of construction for conductors of a constant-current supply circuit involved with a communication circuit and not in Type 1 cable shall be based on either its current rating or on the open-circuit voltage rating of the transformer supplying such circuit, as set forth in Tables 242-1 and 242-2. When the constant current supply circuit is in Type 1 cable, the grade of construction shall be based on its nominal full-load voltage.

B. **Railway Feeder and Trolley-Contact Circuit Conductors**

Railway feeder and trolley contact circuit conductors shall be considered as supply conductors for the purpose of determining the required grade of construction.

C. **Communication Circuit Conductors Used Exclusively in the Operation of Supply Lines**

Communication circuit conductors used exclusively in the operation of supply lines shall have their grade of construction determined as follows:

1. By the requirements for ordinary communication circuits when conforming to Rule 288A3.
2. By the requirements for supply circuits when defined by Rule 288A4.

D. **Fire-Alarm Circuit Conductors**

Fire-alarm circuit conductors shall be considered as other communication circuit conductors except that they shall always meet grade D construction where the span length is from 0 to 150 feet, and grade C construction where the span length exceeds 150 feet.

E. **Neutral Conductors of Supply Circuits**

Supply-circuit neutral conductors, which are effectively grounded throughout their length and are not located above supply conductors of more than 750 volts to ground, shall have the same grade of construction as supply conductors of not more than 750 volts to ground, except that they need not meet any insulation requirements. Other neutral conduc-

Table 242-1. Grades of Construction for Supply Conductors Alone, at Crossing, or on the Same Structures With Other Conductors

(The voltages listed in this table are line to ground values for: effective grounded ac circuits, two wire grounded circuits, or center grounded dc circuits; otherwise line to line values shall be used. The grade of construction for supply conductors, as indicated across the top of the table, must also meet the requirements for any lines at lower levels except when otherwise noted.)

Conductors, tracks and rights of way at lower levels \ Supply conductors at higher levels ①	Constant-potential supply conductors										Constant current supply conductors		Communication conductors used exclusively in the operation of and run as supply lines
	0–0.75 kV		0.75–8.7 kV				Exceeding 8.7 kV						
	Urban	Rural	Urban		Rural		Urban		Rural				
	Open or Cable	Open or Cable	Open	Cable	Open	Cable	Open	Cable	Open	Cable	Open	Cable	Open or cable
Exclusive private rights-of-way	N	N	②N	N	N	N	②N	②N	N	N	B, C, or N; see Rule 242A		C or N; see Rule 242C
Common or public rights-of-way	N	N	C	N	N	N	③C	C	N	N	B, C, or N; see Rule 242A		C or N; see Rule 242C
Railroad tracks and limited access highways	B	B	B	B	B	B	B	B	B	B	B	B	B

220

Table 242-1 *(continued)*

											see Rule 242A	see Rule 242C
Constant potential supply conductors												
0 to 750 V — Open or cable	N	N	C	N	N	N	③C	C	④C	N	B, C, or N; see Rule 242A	B, C, or N; see Rule 242C
750 V to 8.7 kV — Open	⑤C	N	C	C	N	N	③C	C	N		B, C, or N; see Rule 242A	B, C, or N; see Rule 242C
750 V to 8.7 kV — Cable	N	N	C	N	N	N	③C	C	N			
Exceeding 8.7 kV — Open	⑤B	⑤C	B	B	N	N	③C	C	N		B, C, or N; see Rules 242A and 242C	B, C, or N; see Rules 242A and 242C
Exceeding 8.7 kV — Cable	⑤C	N	C	N	N	N	③C	C	N			
Constant current supply conductors: Open or cable	B, C, or N; see Rule 242A										B, C, or N; see Rule 242A	B, C, or N; see Rules 242A and 242C
Communication conductors: Open or cable, used exclusively in the operation of supply lines ⑩	B, C, or N; see Rule 242C										B, C, or N; see Rules 242A and 242C	B, C, or N; see Rule 242C
Communication conductor: Urban or rural, open or cable ⑥	N	N	⑦⑧B	C	⑦⑧B	C	⑧B	C	⑧B	C	⑧⑨B / C or N; see Rule 242A	B, C, or N; see Rule 242C

221

① The words "open" and "cable" appearing in the headings have the following meanings as applied to supply conductors: Cable means the Type 1 cables described in Rule 241A; open means open wire and Type 2 cables.

② Lines that can fall outside the exclusive private rights-of-way shall comply with the grades specified for lines not on exclusive private rights-of-way.

③ Supply conductors shall meet the requirements of grade B construction if the supply circuits will not

(Footnotes for Table 242-1 continued on page 222)

be promptly de-energized, both initially and following subsequent breaker operations, in the event of a contact with lower supply conductors or other grounded objects.

④Grade N construction may be used if crossing over supply services only.

⑤If the wires are service drops, they may have grade N sizes and tensions as set forth in Table 263-2.

⑥Grade N construction may be used where the communication conductors consists only of not more than one insulated twisted-pair or parallel-lay conductor, or where service drops only are involved.

⑦Grade C construction may be used if the voltage does not exceed 2.9 kV.

⑧The supply conductors need only meet the requirements of grade C construction if both of the following conditions are fulfilled:

(1) The supply voltage will be promptly removed from the communication plant by de-energization or other means, both initially and following subsequent circuit breaker operations in the event of a contact with the communication plant.

(2) The voltage and current impressed on the communication plant in the event of a contact with the supply conductors are not in excess of the safe operating limit of the communication protective devices.

⑨Grade C construction may be used if the current cannot exceed 7.5 A or the open-circuit voltage of the transformer supplying the circuit does not exceed 2.9 kV.

⑩Communication circuits located below supply conductors shall not affect the grade of construction of the supply circuits.

Tables 242-2. Grades of Construction for Communication Conductors Alone, or in Upper Position of Crossing or on Joint Poles

(The voltages listed in this table are line to ground values for: effectively grounded ac circuits, two wire grounded circuits, or center grounded dc circuits; otherwise line to line values shall be used. The grade of construction for supply conductors, as indicated across the top of the table, must also meet the requirements for any lines at lower levels except when otherwise noted.)

(Placing of communication conductors at higher levels at crossings, or on jointly used poles should generally be avoided, unless the supply conductors are trolley-contact conductors and their associated feeders.)

Conductors, tracks, and rights-of-way at lower levels	Communication conductors (Communication conductors, rural or urban, open or cable, including communication conductors run as such, but used exclusively in the operation of supply lines.)
Exclusive private right-of-way	N
Common or public rights-of-way	N
Railroad tracks and limited access highways	D
Constant potential supply conductors① 0 to 750 V Open or cable	N
750 V to 2.9 kV Open or cable	C
Exceeding 2.9 kV Open	B
Cable	C
Constant current supply conductors① 0 to 7.5 A Open②	C
Exceeding 7.5 A Open②	③B
Communication conductors, open or cable, used exclusively in the operation of supply lines	④B, C, or N
Communication conductors, open or cable, urban or rural	N

①The words "open" and "cable" appearing in the headlines have the following meaning as applied to supply conductors: Cable means Type 1 cables as described in Rule 241A1; open means open wire and also Type 2 cables, as described in Rule 241A2.

②Where constant current circuits are in Type 1 cable, the grade of construction shall be based on the nominal full-load voltage.

③Grade C construction may be used if the open circuit voltage of the transformer supplying the circuit does not exceed 2.9 kV.

④See Rule 242C.

tors shall have the same grade of construction as the phase conductors of the supply circuits with which they are associated.

F.   Lightning Protection Wires

Lightning protection wires shall be of the same grade of construction as the supply conductors with which they are associated.

243. Grades of Construction for Line Supports

A. Structures

The grade of construction shall be that required for the highest grade of conductors supported except as modified by the following:

1.   The grade of construction of jointly used structures, or structures used only by communication lines, need not be increased merely because the communication wires carried on such structures cross over trolley-contact conductors of 0 to 750 volts to ground.

2.   Structures carrying grade C or D fire-alarm conductors, where alone, or where concerned only with other communication conductors, need meet only the requirements of grade N.

3.   Structures carrying supply service drops of 0 to 750 volts to ground shall have at least the grade of construction required for supply line conductors of the same voltage.

4.   Where the communication lines cross over supply conductors and a railroad in the same span and grade B is required by Rule 241C3b for the communication conductors, due to the presence of railroad tracks, the grade of the structures shall be D.

5.   The grade of construction required for a conflicting structure (first circuit) shall be determined from the requirements of Rule 242 for crossings. The conflicting structure's conductors (first circuit) shall be assumed to cross the other circuit's conductors (second circuit) for the purposes of determining the grade of construction required for the conflicting structure.

NOTE: The resulting structure grade requirement could result in a higher grade of construction for the structure than for the conductors carried thereon.

B. **Crossarms and Support Arms**

The grade of construction shall be that required for the highest grade of conductors carried by the arm concerned except as modified by the following:

1. The grade of construction of arms carrying only communication conductors need not be increased merely because the conductors cross over trolly-contact conductors of 0 to 750 volts to ground.

2. Arms carrying grade C or D fire-alarm conductors, where alone or where concerned with other communication conductors, need meet only the requirements for grade N.

3. Arms carrying supply service drops of 0 to 750 volts to ground shall have at least the grade of construction required for supply line conductors of the same voltage.

4. Where communication lines cross over supply conductors and a railroad in the same span and grade B is required by Rule 241C3b for the communication conductors due to the presence of railroad tracks, the grade of the arm shall be D.

C. **Pins, Armless Construction Brackets, Insulators, and Conductor Fastenings**

The grade of construction for pins and armless construction brackets, insulators, and conductor fastenings shall be that required for the conductor concerned except as modified by the following:

1. The grade of construction need not be increased merely because the supported conductors cross over trolley-contact conductors of 0 to 750 volts to ground.

2. Grade N construction is sufficient when only grade C or D fire-alarm conductors or other communication conductors are concerned.

3. Supply service drops of 0 to 750 volts to ground only require the same grade of construction as supply-line conductors of the same voltage.

4. When grade B construction is required by Rule 241C3b for the communication conductors due to the presence of railroad tracks, grade D construction shall be used when supporting communication lines which cross over supply conductors and a railroad in the same span.

5. When communication conductors are required to meet grade B or C, only the requirements for mechanical strength for these grades is required.

6. Insulators for use on open conductor supply lines shall meet the requirements of Section 27 for all grades of construction.

225

Section 25. Loading for Grades B, C, and D

250. General Loading Requirements and Maps

A. General

1. It is necessary to assume the loadings which may be expected to occur on a line because of wind and ice during all seasons of the year. These minimum weather loadings shall be the values of loading resulting from the application of Rules 250B or 250C. Where both rules apply, the required loading shall be that which, when combined with the appropriate overload capacity factors, has the greater effect on strength requirements.

2. Where construction or maintenance loads exceed those imposed by Rule 250A1, which may occur more frequently in light loading areas, the assumed loadings shall be increased accordingly.

3. It is recognized that loadings actually experienced in certain areas in each of the loading districts may be greater, or in some cases, may be less than those specified in these rules. In the absence of a detailed loading analysis, no reduction in the loadings specified therein shall be made without approval of the administrative authority.

B. Combined Ice and Wind Loading

Three general degrees of loading due to weather conditions are recognized and are designated as heavy, medium, and light loading. Figure 250-1 shows the districts in the states in which these loadings are normally applicable.

NOTE: The localities are classified in the different loading districts according to the relative simultaneous prevalence of wind velocity and thickness of ice which accumulates on wires. Light loading is for places where little, if any, ice accumulates on wires.

Table 250-1 shows the minimum radial thicknesses of ice and the wind pressures to be used in calculating loadings. Ice is assumed to weigh 57 pounds per cubic foot.

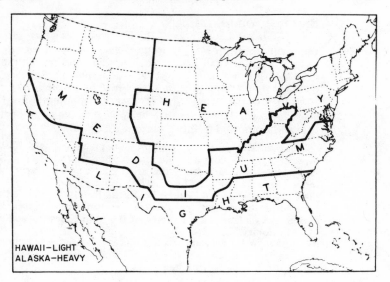

Fig 250-1
**General Loading Map of United States with Respect
to Loading of Overhead Lines**

Table 250-1.   Ice, Wind and Temperature

	Loading Districts (For use with Rule 250B)			Extreme Wind Loading (For use with Rule 250C)
	Heavy	Medium	Light	
Radial thickness of ice (in)	0.50	0.25	0	0
Horizontal wind pressure in lbs per sq ft	4	4	9	See Fig 250-2
Temperature (°F)	0	+15	+30	+60

C. **Extreme Wind Loading**
Figure 250-2 is a wind map of the United States which shows the minimum horizontal wind pressures to be used for calculating loads upon tall structures. For wind pressure at a specific location use a value not less than that of the nearest pressure line. If any portion of a structure of supported facilities is located in excess of 60 feet above ground or water level, these wind pressures shall be applied to the entire structure and supported facilities without ice covering.

NOTE 1: The values of wind pressure given in Figure 250-2 represent the loading of wind upon cylindrical surfaces at 30 feet above ground level. They are based upon 50 year isotachs given in ANSI A58.1-1972, Building Code Requirements for Minimum Design Loads in Buildings and Other Structures, converted from miles per hour to pressure on cylindrical surfaces by the factor of 0.00256 times the square of the velocity.
NOTE 2: Wind velocity usually increases with height; therefore, experience may show that the wind pressures specified herein need to be further increased.

251. Conductor Loading

A. **General**
Ice and wind loads shall be as specified in Rule 250.
1. Where a cable is attached to a messenger, the specified loadings shall be applied to both cable and messenger.
2. In determining wind loadings on a bare stranded conductor or multiconductor cable, the assumed projected area shall be that of a smooth cylinder whose outside diameter is the same as that of the conductor or cable.

   NOTE: Experience has shown that as the size of multiconductor cable decreases, the actual projected area decreases, but the roughness factor increases and offsets the reduction in projected area.

3. In determining loadings on ice-covered bare stranded conductor or multiconductor cables, the coating of ice shall be considered a hollow cylinder touching the outer strands of the bare stranded conductor or the outer circumference of the multiconductor cable. For bundled conductors, the coating of ice shall be considered as individual hollow cylinders around each subconductor.

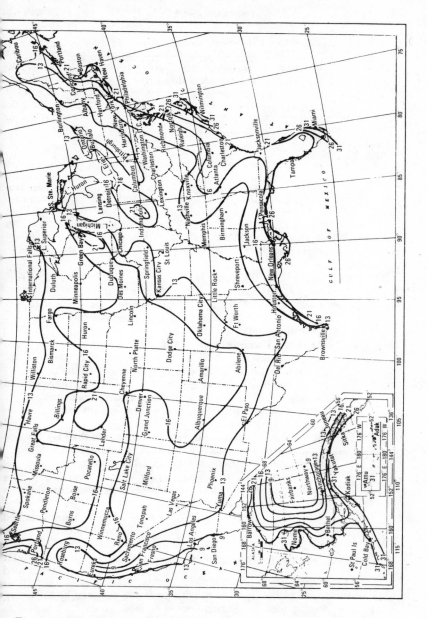

Figure 250-2.   Extreme Wind Pressure in Pounds per Square Foot at
30 ft Above Ground (based on fastest mile of wind)

Table 251-1. Temperatures and Constants

	Loading districts (for use with Rule 250B)			Extreme wind loading (for use with Rule 250C)
	Heavy	Medium	Light	
Temperature (°F)	0	+15	+30	+60
Constant to be added to the resultant in pounds per foot:				
All conductors	0.30	0.20	0.05	0.0

B. Loading Components
   The components of loading and total loading shall be as follows.
   1. Vertical Loading Component
       The vertical load on a conductor or messenger shall be its own weight plus the weight of conductors, spacers, or equipment which it supports, ice covered where specified in Rule 250.
   2. Horizontal Loading Component
       The horizontal load shall be the horizontal wind pressure specified in Rule 250 applied at right angles to the direction of the line to the projected area of the conductor or messenger and conductors, spacers, or equipment which it supports, ice covered where specified in Rule 250.
   3. Total Loading
       The total load on a conductor or messenger shall be the resultant of components 1 and 2 above, calculated at the temperature specified in Table 251-1, to which resultant has been added the constant specified in Table 251-1. In all cases the conductor or messenger tension shall be computed from this total loading.

252. Loads Upon Line Supports
   A. Assumed Vertical Loading

   The vertical loads upon poles, towers, foundations, crossarms, pins, insulators, and conductor fastenings shall be their own weight plus the superimposed weight which they support, including all wires and cables, in accordance with Rules 251A and 251B1, together with the effect of any difference in elevation of supports. The radial thickness of ice shall be computed only upon wires, cables, and messengers, and not upon supports.

B. Assumed Transverse Loading

The total transverse loading upon poles, towers, foundations, crossarms, pins, insulators, and conductor fastenings shall include the following.

1. Transverse Loading from Conductors and Messengers.

The transverse loading from conductors and messengers shall be the horizontal loading specified in Rule 251. For supporting structures carrying more than 10 wires, not including cables supported by messengers, where the pin spacing does not exceed 15 inches, the transverse wind load shall be calculated on two-thirds of the total number of such wires with a minimum of 10 wires, except in light loading areas defined by Rule 250.

2. Structure Loading

The transverse loading upon structures and equipment shall be computed by applying, at right angles to the direction of the line, the appropriate horizontal wind pressure given in Rule 250. This pressure shall be applied upon the projected surfaces of the structures and equipment supported thereon, without ice covering. The following shape factors shall be applied.

a. Cylindrical Structures and Components

Wind loads on straight or tapered cylindrical structures or structures composed of numerous narrow relatively flat panels which combine to form a total cross section that is approximately circular or elliptical in shape shall be computed from the assumed unit wind pressure specified in Rule 250 applied to the projected area multiplied by a shape factor of 1.0.

b. Flat Surfaced Structures and Components

Wind loads on flat surfaced structures, having solid or enclosed flat sides and an overall cross section that is substantially square or rectangular, shall be computed from the assumed unit wind pressures specified in Rule 250 applied to the projected area of one face multiplied by a shape factor of 1.6 to allow for pressure on flat surfaces.

c. Latticed Structures

Wind loads on essentially square or rectangular latticed structures or components shall be computed from the assumed unit wind pressures specified in Rule 250 applied to the sum of the projected areas of the members of the front face multiplied by a shape factor of 3.2 to allow for wind pressure if structural members are flat surfaced or 2.0 if structural surfaces are cylindrical. The total, however, need not exceed the load which would occur on a solid structure of the same outside dimension.

EXCEPTION: The shape factors listed under Rules 252B2a, 252B2b, and 252B2c may be reduced if wind tunnel tests or rational aerodynamic analysis produce evidence that such a reduction is justifiable. In the absence of such tests or analyses, the factors given above shall be considered to be minimum values.

3. At Angles

Where a change in direction of wires occurs, the loading upon the structure, including guys, shall be assumed to be a resultant load equal to the vector sum of the transverse wind load as derived above and the resultant load imposed by the wires due to their change in direction. In deriving these loadings, a wind direction shall be assumed which will give the maximum resultant load, proper reduction being made in loading to account for the reduced wind pressure on the wires resulting from the angularity of the application of the wind to the wires.

4. Span Lengths

The calculated transverse load shall be based upon the average of the actual lengths of the two spans adjacent to the structure concerned.

C. Assumed Longitudinal Loading

1. Change in Grade of Construction

The longitudinal loading upon supporting structures, including poles, towers, and guys at the ends of sections required to be of grade B construction, when located in lines of lower than grade B construction, shall be taken as an unbalanced pull in the direction of the higher grade section equal to the larger of the following values:

a. The pull of two-thirds, and in no case less than two of the conductors which have rated breaking strength of

3000 pounds or less, such two-thirds of the conductors being selected so as to produce the maximum stress in the support.

b. The pull of one conductor when there are eight or less conductors (including overhead ground wires) having rated breaking strength of more than 3000 pounds, and the pull of two conductors when there are more than eight conductors, such conductors being selected so as to produce the maximum stress in the support.

2. Jointly Used Poles at Crossings Over Railroads, Communication Lines, or Limited Access Highways
Where a joint line crosses over a railroad, a communication line, or a limited access highway, and grade B is required for the crossing span, the tension in the communication conductors of the joint line shall be considered as limited to one-half their rated breaking strength, provided they are smaller than No. 8 Stl. WG, if of steel, or No. 6 AWG, if of copper.

3. Dead Ends
The longitudinal loading upon supporting structures at dead ends for line terminations shall be taken as an unbalanced pull equal to the tensions of all conductors and messengers (including overhead ground wires); except that with spans in each direction from the dead-end structure, the unbalanced pull shall be taken as the difference in tensions.

4. Unequal Spans and Unequal Vertical Loading
Where longitudinal loads can be created by the difference in tensions in the wires in adjacent spans caused by unequal vertical loading or unequal spans, the structures should be capable of supporting this unbalanced longitudinal loading.

5. Stringing Loads
Proper allowance should be made for longitudinal loads which may be produced on the structures by wire stringing operations.

6. Longitudinal Capability
It is recommended that structures having a longitudinal strength capability be provided at reasonable intervals along the line.

7. Communication Conductors on Unguyed Supports at Railroad Crossings and Limited Access Highways

The longitudinal loading shall be assumed equal to an unbalanced pull in the direction of the crossing of all open-wire conductors supported, the pull of each conductor being taken as 50 percent of its rated breaking strength in the heavy loading district, $33\frac{1}{3}$ percent in the medium loading district, and $22\frac{1}{4}$ percent in the light loading district.

D. Simultaneous Application of Loads
Where a combination of vertical, transverse, or longitudinal loads may occur simultaneously, the structure shall be designed to withstand the simultaneous application of these loads.

NOTE: Under the extreme wind conditions of Rule 250C, an oblique wind may require greater structural strength than that computed under Rules 252B and C.

Section 26. Strength Requirements

260. Preliminary Assumptions (see also Section 20)

A. It is recognized that deformation, deflections, or displacement of parts of the structure will, in some cases, change the effects of the loads assumed. In the calculation of stresses, allowance may be made for such deformation, deflection or displacement of supporting structures including poles, towers, guys, crossarms, pins, conductor fastenings, and insulators when the effects can be accurately evaluated. Such deformation, deflection, or displacement should be calculated using the Rule 250 loads prior to application of the overload factors required by this section. For crossings or conflicts, the calculations shall be subject to mutual agreement.

B. It is recognized that newly developed materials may become available. It is further recognized that, while these materials are in the process of development, they must be tested and evaluated. Trial installations are permitted where qualified supervision is provided.

C. The overload capacity factors shown in the tables of this section apply for the combined ice and wind loading conditions specified in Rule 250B. For the extreme wind loading condition specified in Rule 250C, an overload capacity factor of not less than 1.0 shall be applied for structures and their foundations, and 1.25 for other supported facilities.

261. Grades B and C Construction

A. Supporting Structure

The strength requirements for supporting structures may be met by the structures alone or with the aid of guys and/or braces.

1. Metal, Prestressed, and Reinforced Concrete Structures
The structures shall be designed to withstand the loads in Rule 252 multiplied by the appropriate overload capacity factors given in Tables 261-1 or 261-2. (Where guys are used, see Rule 261C.)

a. Minimum Strength
All structures (including those below 60 feet) shall withstand, without conductors, the extreme wind pressure in Rule 252 applied in any direction on the structure times an overload capacity factor of 1.0. A gust factor appropriate for the wind pressure and structure height should be considered.

b. Strength at Angles in a Line
At an angle in a line, the strength of the support shall be sufficient to withstand the total transverse loadings specified in Rule 252 multiplied by the appropriate overload capacity factor for transverse strength given in Tables 261-1 or 261-2.

2. Wood Structures
Wood structures shall be of such material and dimensions as to meet the following requirements. (Where guys are used, see Rule 261C):

a. Designated Fiber Stress
(1) Natural wood poles of various species meeting the requirements of ANSI 05.1-1979, Specifications and Dimensions for Wood Poles, shall be considered as having the designated fiber stresses set forth in that standard.
(2) Appropriate adjustments in designated fiber stresses shall be made for sawn or laminated wood.

Table 261-1. Overload Capacity Factors for Reinforced Concrete Structures (Not Prestressed)

	Overload capacity factors	
	Grade B	Grade C
Vertical strength	4.0	2.67
Transverse strength		
Wind load	4.0	2.67
Wire tension load at angles	2.0	1.33
Longitudinal strength		
In general	1.0	no requirement
At dead ends	2.0	1.33

NOTE: The factors in this table apply for the loading conditions of Rule 250B. For extreme wind loading conditions see Rule 260C.

Table 261-2. Overload Capacity Factors for Metal and Prestressed Concrete Structures

	Overload capacity factors	
	Grade B	Grade C
Vertical strength	1.50	1.10
Transverse strength		
Wind load	2.50	2.20
Wire tension load at angles	1.65	1.10
Longitudinal strength		
At Crossings		
In general	1.10	no requirement
At dead ends	1.65	1.10
Elsewhere		
In general	1.00	no requirement
At dead ends	1.65	1.10

NOTE: The factors in this table apply for the loading conditions of Rule 250B. For extreme wind loading conditions, see Rule 260C.

b. Transverse and Vertical Strength
    Wood structures shall be designed to withstand the
    transverse and vertical loads in Rule 252, multiplied
    by the appropriate overload capacity factor given in
    Table 261-3, without exceeding the designated fiber
    stress.

    EXCEPTION: When installed, naturally grown wood
    poles acting as single based structures or unbraced
    multiple pole structures, shall meet the requirements
    of Rules 261A2b (261A2c) without exceeding the
    designated fiber stress at the ground line for unguyed
    poles or at the point of attachment for guyed poles.

c. Longitudinal and Dead-End Strength
    Wood structures shall be designed to withstand the
    longitudinal and dead-end loadings in Rule 252
    multiplied by the appropriate overload capacity fac-
    tor in Table 261-3 without exceeding the designated
    fiber stress.

    EXCEPTION 1: At a grade B crossing, in a straight
    section of line, wood structures complying with the
    transverse strength requirements of Rule 261A2b,
    without the use of transverse guys shall be considered
    as having the required longitudinal strength, providing
    the longitudinal strength is comparable to the trans-
    verse strength of the structure. This exception does
    not modify the requirements of this rule for dead ends.

    EXCEPTION 2: At a grade B crossing of a supply
    line over a highway or a communication line where
    there is an angle in the supply line, wood structures
    shall be considered as having the required longitudinal
    strength if all of the following conditions are met:

    (1) The angle is not over 20 degrees.
    (2) The angle structure is guyed in the plane of the
         resultant of the conductor tensions. The tension
         in this guy under the loading in Rule 252
         multiplied by an overload capacity factor of 2.0
         shall not exceed the allowable guy value specified
         in Rule 261C.
    (3) The angle structure has sufficient strength to
         withstand, without guys, the transverse loading of
         Rule 252, which would exist if there were no
         angle at that structure with an overload capacity
         factor of 4.0 when installed or 2.67 at replace-
         ment.

EXCEPTION 3: When installed, naturally grown wood poles acting as single based structures or unbraced multiple pole structures, shall meet the requirements of Rules 261A2b (261A2c) without exceeding the designated fiber stress at the ground line for unguyed poles or at the point of attachment for guyed poles.

d. Strength at Angles in a Line

At an angle in the line, the wood structure shall be designed to withstand the total transverse loading in Rule 252 multiplied by the appropriate overload capacity factor given in Table 261-3 without exceeding the designated fiber stress.

Table 261-3. Overload Capacity Factors for Wood Structures

	Grade B		Grade C	
	When installed	At replacement	When installed	At replacement
Transverse (wind) and Vertical strength				
At Crossings	4.0	2.67	2.67	1.33
Elsewhere	4.0	2.67	2.00	1.33
Transverse (wire tension load) strength				
At Crossings	2.0	1.33	1.33	1.00
Elsewhere	2.0	1.33	1.33	1.00
Longitudinal Strength				
In general	1.33	1.00	no requirement	no requirement
At dead-ends	2.00	1.33	1.33	1.00

NOTES: (1) Where structures are built for temporary service the overload capacity factors at replacement may be used provided that the designated fiber stress is not exceeded during the life of the structure.

(2) The factors in this table apply for the loading conditions of Rule 250B. For extreme wind loading conditions, see Rule 260C.

e. Strength of Guyed Poles

Guyed poles shall be designed as columns, resisting the vertical component of the tension in the guy plus any other vertical loads on such poles.

f. Spliced and Stub-Reinforced Poles

The use of stub reinforcements or permanent splices at any section along the pole that develop the required strength of the pole is permitted, provided the remainder of the pole is in good condition and is of sufficient size to develop its required strength.

g. Average Strength of Three Poles

A pole (single-base structure) not individually meeting the transverse strength requirements will be permitted when reinforced by a stronger pole on each side, if the average strength of the three poles meets the transverse strength requirements, and the weak pole has not less than 75 percent of the required strength. An extra pole inserted in a normal span for the purpose of supporting a service drop may be ignored.

EXCEPTION: This rule does not apply to crossings over railroads, communication lines, or limited access highways.

3. Transverse-Strength Requirements for Structures Where Side Guying Is Required, But Can Only Be Installed at a Distance

Grade B: In the case of structures where, because of very heavy or numerous conductors or relatively long spans, the transverse-strength requirements of this section cannot be met except by the use of side guys or special structures, and if it is physically impractical to employ side guys, the transverse-strength requirements may be met by side-guying the line at each side of, and as near as practical to, the crossing, or other transversely weak structure, and with a distance between such side-guyed structures of not over 800 feet, provided that:

a. The side guyed structures for each such section of 800 feet or less shall be constructed to withstand the calculated transverse load due to wind on the supports and ice-covered conductors, on the entire section between the side-guyed structures.

b. The line between such side-guyed structures shall be substantially in a straight line and the average length of span between the side guyed structures shall not exceed 150 feet.

c. The entire section between the transversely strong structures shall comply with the highest grade of construction concerned in the given section, except as to the transverse strength of the intermediate poles or towers.

Grade C: The above provisions do not apply to grade C.

4. Longitudinal-Strength Requirements for Sections of Higher Grade in Lines of a Lower Grade Construction

a. Methods of Providing Longitudinal Strength

Grade B: The longitudinal-strength requirements for sections of line of higher grade in lines of a lower grade (for assumed longitudinal loading, see Rule 252) may be met by placing supporting structures of the required longitudinal strength at either end of the higher grade section of the line.

Where this is impractical, the supporting structures of the required longitudinal strength may be located one or more span lengths away from the section of higher grade, within 500 feet on either side and with not more than 800 feet between the longitudinally strong structures, provided such structures and the line between them meet the requirements as to transverse strength and stringing of conductors, of the highest grade occurring in the section, and provided that the line between the longitudinally strong structure is approximately straight or suitably guyed.

The requirements may also be met by distributing the head guys over two or more structures on either side of the crossing, such structures and the line between them complying with the requirements for the crossing as to transverse strength and as to conductors and their fastenings. Where it is impractical to provide the longitudinal strength, the longitudinal loads shall be reduced by increasing the conductor sags. This may require greater conductor separations. (See Rule 235 B.)

Grade C: The above provisions do not apply to grade C.

b. Flexible Supports

Grade B: When supports of the section of higher grade are capable of considerable deflection in the direction of the line, as with wood or concrete poles, or some types of metal poles and towers, it may be necessary to increase the normal clearances specified in Section 23, or to provide head guys or special reinforcement to prevent such deflection.

Flexible metal structures may have to be head-guyed or otherwise reinforced to prevent reduction in the clearances required in Section 23.

Grade C: The above provision does not apply to grade C.

B. Strength of Foundations and Settings

The loadings in Rule 252 multiplied by the overload factors given in Table 261-4 shall be applied to the structure. Foundations and settings shall be designed or be determined by experience to withstand the reactions resulting from these applied loadings.

NOTE: Excessive movement of foundations and guy anchors may reduce structure strength or impair clearances.

Table 261-4. Overload Capacity Factors for
Foundations and Settings

| | Overload capacity factors | |
	Grade B	Grade C
Vertical strength	1.5	1.1
Transverse strength		
Wind load	2.5	2.2
Wire tension load	1.65	1.1
Longitudinal strength		
In general	1.1	1.0
At dead ends	1.65	1.1

NOTE: The factors in this table apply for the loading conditions of Rule 250B. For extreme wind loading conditions, see Rule 260C.

C. **Strength of Guys and Guy Anchors**
The general requirements for guys and guy insulators are
covered under Rules 282 and 283, respectively. Guy anchors
shall withstand the loads in Rule 252 multiplied by the
overload factors given in Table 261-5.
1. Metal and Prestressed Concrete Structures
    Guys shall be considered as an integral part of the
    structure and shall withstand the loads in Rule 252,
    multiplied by the overload factors given in Table 261-2,
    without exceeding 90 percent of the rated breaking
    strength of the guy.
2. Wood and Reinforced Concrete Poles and Structures
    When guys are used to meet the strength requirements
    they shall be considered as taking the entire load in the
    direction in which they act, the structure acting as a strut
    only, except for those structures considered to possess
    sufficient rigidity so that the guy can be considered an
    integral part of the structure.
    a. Guys shall be of such material and dimension to
        withstand the loads in Rule 252, multiplied by the
        overload capacity factors given in Table 261-5 with-
        out exceeding 90 percent of the rated breaking
        strength of the guy.
    b. At an angle in the line, the guy shall be of such
        material and dimension to withstand the total trans-

Table 261-5. Overload Capacity Factors for Guys

	Overload capacity factors	
	Grade B	Grade C
Transverse strength		
Wind load	2.67	2.0
Wire tension load	1.5	1.15
Longitudinal strength (except at angles)		
In general	1.0	no requirement
At dead ends	①1.5	①1.15

① If deflection of supporting structures is taken into account in the
computations, the overload capacity factors of 1.5 shall be increased
to 1.67; 1.15 shall be increased to 1.33.
NOTE: The factors in the table apply for the loading conditions
of Rule 250B. For extreme wind loading conditions, see Rule 260C.

verse loads in Rule 252, multiplied by the overload capacity factors given in Table 261-5, without exceeding 90 percent of the rated breaking strength of the guy.

D. Crossarms

1. Vertical Strength

Crossarms shall withstand the vertical loads specified in Rule 252 without exceeding 50 percent of the designated fiber stress of the material (or ultimate strength) where applicable.

2. Bracing

Crossarms shall be securely supported by bracing, if necessary, so as to support safely all expected loads to which they may be subjected in use including line personnel working on them.

3. Longitudinal Strength

a. General

Crossarms shall withstand without exceeding their designated fiber stress (or ultimate strength), the applicable longitudinal loads given in Rule 252, or 700 pounds applied at the outer conductor attachment points, whichever is greater. At each end of a transversely weak section, as described in Rule 261A3, the longitudinal load shall be applied in the direction of the weak section.

b. Methods of Meeting Rule 261D3

Grade B: Where conductor tensions are limited to a maximum of 2000 pounds per conductor, double wood crossarms having cross sections specified in Table 261-6 and properly assembled, will be considered as meeting the strength requirements specified in Rule 261D3a.

Grade C: This requirement is not applicable.

4. Material and Minimum Size

Wood crossarms of selected southern pine or douglas fir shall have a cross section of not less than those shown in Table 261-6. Crossarms of other suitable timber or of other materials may be used provided they are of equivalent strength.

5. Double Crossarms or Brackets

Grade B: Where pin type construction is used, double crossarms or support of equivalent strength shall be used at each crossing structure, at ends of joint use or conflict

**Table 261-6. Minimum Dimensions of
Crossarm Cross Section in Inches**

Number of pins	Grades of construction		
	Grade B	Grade C	
		Supply	Communication
2 or 4	3 by 4	2¾ by 3¾	—
6 or 8	3¼ by 4¼	3 by 4	—
6	—	—	2¾ by 3¾
10	—	—	3 by 4

sections, at dead ends and at corners where the angle of departure from a straight line exceeds 20 degrees. Under similar conditions, where a bracket supports a conductor operated at more than 750 volts to ground and there is no crossarm below, double brackets shall be used.

EXCEPTION: The above does not apply where communication cables or conductors cross below supply conductors and either are attached to the same pole, or where supply conductors are continuous and of uniform tension in the crossing span and each adjacent span. This exception does not apply to railroad crossings and limited access highways except by mutual agreement. Grade C: The above requirement is not applicable.

    6. Location

At crossing, crossarms should be attached to the face of the structure away from the crossing, unless special bracing or double crossarms are used.

E. Metal Crossarms

Metal crossarms shall withstand the loads in Rule 252 multiplied by the overload capacity factors in Table 261-2.

F. Strength of Pin Type or Similar Construction and Conductor Fastenings.

    1. Longitudinal Strength

      a. General

Pin type or similar construction and ties or other conductor fastenings shall withstand the applicable longitudinal loads given in Rule 252, or 700 pounds applied at the pin, whichever is greater. At each end of a transversely weak section as described in Rule 261A3, the longitudinal load shall be applied in the direction of the weak section.

Grade C: No requirement.

b. Method of Meeting Rules 261F1a.

Grade B: Where conductor tensions are limited to 2000 pounds and such conductors are supported on pin insulators, double wood pins and ties or their equivalent, will be considered to meet the requirements of Rule 261F1a.

Grade C: No requirement.

c. At Dead Ends and at Ends of Higher Grade Construction in Line of Lower Grade

Grade B: Pins and ties or other conductor fastenings connected to the structure at a dead end or at each end of the higher grade section shall be of sufficient strength to withstand at all times without exceeding their ultimate strength, an unbalanced pull due to the conductor loading specified in Rule 251.

Grade C: This requirement is not applicable except for dead ends.

d. At Ends of Transversely Weak Sections

Grade B: Pins and ties or other conductor fastenings connected to the structure at each end of the transversely weak section as described in Rule 261A3 shall be such as to withstand at all times without exceeding their ultimate strength, the unbalanced pull in the direction of the transversely weak section of the conductor supported, under the loading prescribed in Rule 251.

Grade C: No requirement.

2. Double Pins and Conductor Fastenings

Grade B: Where wood pins are used, double pins and conductor fastenings shall be used where double crossarms or brackets are required by Rule 261D5.

EXCEPTION: The above does not apply where communication cables or conductors cross below supply conductors and either are attached to the same pole, or where supply conductors are continuous and of uniform tension in a crossing span and each adjacent span. This exception does not apply in the case of railroad crossings and limited access highway crossing except by mutual agreement.

Grade C: No requirement.
3. Single Supports Used in Lieu of Double Wood Pins.
A single conductor support and its conductor fastening when used in lieu of double wood pins shall develop strength equivalent to double wood pins and their conductor fastenings as specified in Rule 261F1a.

G. Armless Construction.
1. General.
Open conductor armless construction is a type of open conductor supply line construction in which conductors are individually supported at the structure without the use of crossarms.
2. Insulating Material.
Strength of insulating material shall meet the requirements of Section 27.
3. Other Components.
Strengths of other components shall meet the appropriate requirements of Rules 260 and 261.

H. Open Supply Conductors
1. Minimum Sizes of Supply Conductors
Supply conductors shall have a rated breaking strength and an overall diameter of metallic conductor not less than that of medium-hard-drawn copper of the AWG size shown in Table 261-7 except that conductors made entirely of bare or galvanized iron or steel shall have an overall diameter not less than Stl. WG of the gage sizes shown.

EXCEPTION 1: At railroad crossings, for stranded conductors, other than those in which a central core is entirely covered by the outside wires, any individual wire of such a stranded conductor containing steel shall be not less than 0.100 inch in diameter if copper or aluminum clad and not less than 0.115 inch in diameter if otherwise protected or if bare.

EXCEPTION 2: Service drops of 0 to 750 volts to ground may have the sizes set forth in Rule 263E.

2. Sags and Tensions.
Conductor sags shall be such that, under the assumed loading of Rule 251 for the district concerned, the tensions of the conductor shall not be more than 60 percent of its rated breaking strength. Also the tension at $60°F$, without external load, shall not exceed the following

Table 261-7. Minimum Conductor Sizes

Grade of Construction	Gage Size[1]
B	6
C	8

[1] For No. 6 and No. 8 medium-hard-drawn copper wire, the nominal diameters are 0.1620 and 0.1285 in and the minimum values of breaking load are 1010 and 643.9 lbs, respectively. For steel wire gage, the nominal diameters are 0.192 in for No. 6 and 0.162 in for No. 8.

percentages of the conductor rated breaking strength:

    Initial unloaded tension   35 percent
    Final unloaded tension   25 percent

EXCEPTION: In the case of conductors having a cross-section of a generally triangular shape, such as cables composed of three wires, the final unloaded tension at $60°$ F shall not exceed 30 percent of the rated breaking strength of the conductor.

NOTE 1: The above limitations are based on the use of recognized methods for avoiding fatigue failures by minimizing chafing and stress concentration. If such practices are not followed, lower tensions should be employed.

NOTE 2: The factors listed above apply for the loading conditions of Rule 250B. For extreme wind loading conditions, See Rule 260C.

3. Splices, Taps, and Dead-End Fittings

    a. Splices should be avoided in crossings and adjacent spans. If it is impractical to avoid such splices, they shall be of such a type and so made as to have a strength substantially equal to that of the conductor on which they are placed.

    b. Taps should be avoided in crossing spans but if required shall be of a type which will not impair the strength of the conductors to which they are attached.

    c. Dead-end fittings, including the attachment hardware, shall have sufficient strength to withstand the maximum tension resulting from the loads in Rule 251 multiplied by an overload factor of 1.65.

    4.   Trolley-Contact Conductors
In order to provide for wear, no trolley-contact conductor shall be installed of less size than No. 0 AWG, if of copper, or No. 4 AWG, if of silicon bronze.

I.   Supply Cable Messengers

    1.   Messengers shall be stranded and shall not be stressed beyond 60 percent of their rated breaking strength under the loadings specified in Rule 251.

NOTE 1:   There are no strength requirements for cables supported by messengers.
NOTE 2:   Bonding and grounding requirements for Type 1 supply cables are in Section 21.
NOTE 3:   The factor in Rule 261I1 applies for the loading conditions of Rule 251, except when the extreme wind loading conditions, Rule 260C, apply.

J.   Open-Wire Communication Conductors

Open-wire communication conductors in grade B or C construction shall have the sizes and sags given in Rules 261H1 and 261H2 for supply conductors of the same grade.

EXCEPTION: Where open-wire communication conductors in spans of 150 feet or less are above supply circuits of 5 kilovolts or less between conductors, grade C sizes and sags may be replaced by grade D sizes and sags, except that where the supply conductors are trolley-contact conductors of 0 to 750 volts to ground, No. 12 steel wire may be used for spans of 0 to 100 feet, and No. 9 steel wire may be used for spans of 125 to 150 feet.

K.   Communication Cables

    1.   Communication Cables
There are no strength requirements for such cables supported by messengers.

    2.   Messenger
The messenger shall not be stressed beyond 60 percent of its rated breaking strength under the loadings specified in Rule 251.

L.   Paired Communication Conductors

    1.   Paired Conductors Supported on Messenger
       a. Use of Messenger
A messenger may be used for supporting paired conductors in any location, but is only required for paired conductors crossing over trolley-contact conductors of more than 7.5 kilovolts to ground.

b. Sag of Messenger

Messenger used for supporting paired conductors required to meet grade B construction because of crossing over trolley-contact conductors shall meet the sag requirements for grade D messengers.

c. Size and Sag of Conductors

There are no requirements for paired conductors when supported on messenger.

2. Paired Conductors Not Supported on Messenger

a. Above Supply Lines

Grade B: Sizes and sags shall be not less than those required by Rules 261H1 and 261H2 for supply conductors of similar grade.

Grade C: Sizes and sags shall be not less than the following:

Spans 0 to 100 feet—No sag requirements. Each conductor shall have a rated breaking strength of not less than 170 lbs.

Spans 100 to 150 feet—Sizes and sags shall be not less than required for grade D communication conductors.

Spans exceeding 150 feet—Sizes and sags shall be not less than required for grade C supply conductors. (see Rule 261H2.)

b. Above Trolley-Contact Conductors

Grade B: Sizes and sags shall be not less than the following:

Spans 0 to 100 feet—No size requirements. Sags shall be not less than for No. 8 AWG hard-drawn copper. (See Rule 261H2.)

Spans exceeding 100 feet—Each conductor shall have a rated breaking strength of not less than 170 pounds. Sags shall be not less than for No. 8 AWG hard-drawn copper. (See Rule 261H2.)

Grade C: Sizes and sags shall be as follows:

Spans 0 to 100 feet—No requirements.

Spans exceeding 100 feet—No sag requirements. Each conductor shall have a rated breaking strength of not less than 170 pounds.

262. Grade D Construction

A. Poles

1. Designated Fiber Stress

   Natural wood poles of various species meeting the requirements of ANSI 05.1-1979 shall be considered as having the designated fiber stresses set forth in that standard.

2. Strength of Unguyed Poles

   Unguyed poles shall withstand the vertical and transverse loads in Rules 252A and 252B, and the longitudinal loads in Rule 252C7, multiplied by the overload capacity factors given in Table 262-1 without exceeding the designated fiber stress.

3. Strength of Guyed Poles

   Guyed poles shall be designed as columns, resisting the vertical component of the tension in the guy plus any other vertical loads on such poles.

4. Spliced and Stub-Reinforced Poles

   The use of stub reinforcements or permanent splices at any section along the pole that develop the required strength of the pole is permitted, provided the remainder of the pole is in good condition and is of sufficient size to develop its required strength.

B. Pole Settings

   Foundations and settings for unguyed poles shall be such as to withstand the loads assumed in Rules 252A, 252B, and 252C.

Table 262-1. Overload Capacity Factors for Unguyed Wood Poles

	Overload capacity factors
Vertical and transverse strength	
When installed	4.0
At replacement	2.67
Longitudinal Strength	
When installed	1.33
At replacement	1.0

NOTE: The factors in this table apply for the loading conditions of Rule 250B. For extreme wind loading conditions, see Rule 260C.

C. Guys
1. General
The general requirements for guys are covered in Rules 282 and 283.
2. Side Guys
a. Side guys or braces shall be installed on poles supporting the crossing span where required to withstand the loads specified in Rule 252.

EXCEPTION 1: Side guys are not required where the crossing poles have the transverse strength specified in Rule 262A2 without the reduction for conductor shielding otherwise allowed in Rule 252B1.

EXCEPTION 2: Where a line crossing a railroad or highway changes direction more than 10 degrees at either crossing support, the side guy within the angle may be omitted.

EXCEPTION 3: This rule does not apply to crossing poles under the special conditions set forth in Rule 262C5.

3. Longitudinal Guys
Longitudinal (head) guys shall be provided where required to meet the longitudinal strength requirements of Rule 252.

EXCEPTION: Longitudinal guys are not required where the crossing poles have the longitudinal strength specified in Rule 262A2, or for lines carrying only aerial cable. For lines carrying both open wire and aerial cable, head guying is required only for the number of open wires in excess of 10 if the cable is supported by a 6000 pound messenger, or for the number of open wires in excess of 20 if the cable is supported by a 10 000 pound or stronger messenger.

4. Strength of Guys
a. Guys shall be of such material and dimensions to withstand the transverse and longitudinal loads in Rule 252, multiplied by the overload capacity factors given in Table 262-2, without exceeding 90 percent of their rated breaking strength.
b. At an angle in the line, the guy shall be of such material and dimension to withstand the total transverse loads in Rule 252, multiplied by the overload capacity factors given in Table 262-3 without exceeding 90 percent of the rated breaking strength of the guy.

Table 262-2. Overload Capacity Factors for Guys

	Overload capacity factors
Transverse strength	2.67
Longitudinal strength	
In general	1.0
At dead ends	1.5

NOTE: The factors in the table apply for the loading conditions of Rule 250B. For extreme wind loading conditions, see Rule 260C.

Table 262-3. Overload Capacity Factors for Guys at Angles in the Line

	Overload capacity factors Grade B
Transverse strength	
Wind load	2.67
Wire tension load	1.5

5. **Where Guying Is Required But Cannot Be Installed on the Crossing Pole**

When the transverse-strength requirements cannot be met except by side guys and it is physically impractical to employ side guys, the transverse-strength requirements may be met by side-guying the line at each side of, and as near as is practical to, the crossing or other transversely weak structure, and with a distance between such side-guyed structures of not over 800 feet, provided that:

a. The side-guyed structures for each such section of 800 feet or less shall be constructed to withstand the calculated transverse load due to wind on the supports and ice covered conductors, on the entire section between the side-guyed structures.

b. The line between such side-guyed structures shall be substantially in a straight line and the average length of span between the side-guyed structures shall not exceed 150 feet.

**Table 262-4. Minimum Dimensions of Crossarm
Cross Sections**

Maximum number of wires to be carried	Nominal length (ft)	(in)	Cross section (in)
2	1	4½	2⁵⁄₁₆ by 3⁵⁄₁₆
4	3	4½	2⁵⁄₁₆ by 3⁵⁄₁₆
6	6	0	2¾ by 3¾
10	8	6	2¾ by 3¾
10	10	0	3 by 4
①12	10	0	3¼ by 4¼
②16	10	0	3¼ by 4¼

① Where crossarms are bored for ½ in steel pins, 3 in by 4½ in crossarms may be used.
② Permitted in medium-and light-loading districts only.

c. The entire section between the transversely strong structures shall comply with the highest grade of construction concerned in the given section, except as to the transverse strength of the intermediate structures.

D. Crossarms
1. Material and Minimum Size
Wood crossarms of Southern pine or Douglas fir supporting the crossing span shall have a cross section not less than those shown in Table 262-4. Crossarms of other suitable timber or of other materials may be used provided they are of equivalent strength.
2. Double Crossarms
Double crossarms or a support of equivalent strength shall be used at each crossing pole.
EXCEPTION: Single dead-end type crossarms may be used where it is necessary to dead-end conductors of the crossing span, provided such crossarms and associated dead-end fastenings are of sufficient size and strength to withstand the maximum tension of the conductors under the loading specified in Rule 251 and provided further that the conductors are dead-ended on insulators so designed and installed that the conductor will not fall in the event of insulator breakage.

E. Brackets and Racks

Wood brackets may be used only if used in duplicate or otherwise designed so as to afford two points of support for each conductor. Single metal brackets, racks, drive hooks or other fixtures may be used if designed and attached in such manner as to withstand the full dead-end pull of the wires supported.

F. Pins.

1. Strength

Insulator pins shall have sufficient strength to withstand all expected loads to which they may be subjected.

2. Size

a. Wood pins

Wood pins shall be sound and straight grained with a diameter of shank not less than 1¼ inches.

b. Metal pins

Steel or iron pins shall have diameters of shank not less than ½ inch.

G. Insulators.

Each insulator shall be of such pattern, design, and material that when mounted it will withstand without injury and without being pulled off the pin, all expected loads to which they may be subjected.

H. Conductors.

1. Size

Conductors of the crossing span, if of hard-drawn copper or galvanized steel, shall have sizes not less than given in the specifications a and b that follow. Conductors of material other than the above shall be of such size and so strung as to have a mechanical strength not less than that of the sizes of copper conductors given in specifications a and b that follow.

a. Ordinary Span Lengths.

The sizes in Table 262-5 apply.

b. Long Spans

If spans in excess of those specified in Table 262-5 are necessary, the size of conductors shall be increased so that the stress in the conductor will not exceed the limitations of Rule 262H3.

2. Paired Conductors Without Messengers.

Paired wires without a supporting messenger shall be eliminated as far as practical but where used shall meet

Table 262-5.   Minimum Wire Sizes
With Respect to Loading District and Span Length

	Spans (ft)	
Heavy-loading district	0–125	126–150
Medium-loading district	0–150	151–175
Light-loading district	0–175	176–200
	Minimum wire sizes	
Copper, hard-drawn (AWG)	10	9
Steel, galvanized (steel WG)		
In general	10	8
In rural districts of arid regions	12	10
Aluminum or copper clad steel (AWG)	10	9

the following requirements.

a. Strength

Each conductor shall have a rated breaking strength of 170 pounds.

b. Limiting span lengths

Paired wires shall not be used without a supporting messenger in spans longer than 100 feet in the heavy loading district, 125 feet in the medium loading district, and 150 feet in the light loading district.

3. Sags

Conductor sags shall be such that, under the assumed loading or Rule 251 for the district concerned, and assuming rigid structures for the purpose of calculations, the tension of the conductor shall not be more than 60 percent of its rated breaking strength. Also the final unloaded tensions at 60° F, shall not exceed 25 percent of the conductor rated breaking strength.

NOTE: The factors in Rule 262H3 apply for the loading conditions of Rule 250B. For extreme wind loading conditions see Rule 260C.

4. Splices and Taps

Splices shall, as far as practical, be avoided in the crossing and adjacent spans. If it is impractical to avoid such splices, they shall be of such type and so made as to have a strength substantially equal to that of the conductor in which they are placed.

Taps shall be avoided in the crossing span where practical, but if required shall be of a type which will not impair the strength of the conductors to which they are attached.

I. Messengers.
1. Minimum size
Messengers shall be stranded material with a rated breaking strength of 6000 pounds.
2. Sags and Tensions
Multiple-conductor cables and their messengers shall be so suspended that when they are subjected to the loading prescribed in Rule 251, the tension in the messenger shall not exceed 60 percent of its rated breaking strength.
NOTE: The factor in Rule 262I2 applies for the loading conditions of Rule 251, except for extreme wind-loading conditions where Rule 260C applies.

263. Grade N Construction
A. Poles
Poles used for lines for which neither grade B, C, or D is required shall be of such initial size and so guyed or braced, where necessary, as to withstand all expected loads to which they may be subjected, including line personnel working on them. Such poles and stubs on highways shall be located as far as is practical from the traveled portion of highways. The number of crossings over highways should be kept to a minimum. Such poles and stubs located within falling distance of the traveled way of highways, or so located that their failure would permit wires, cables, guys, or other equipment to fall into the traveled way of the highway, or would reduce the clearances specified in Table 232-1 over the highway, shall be periodically inspected and maintained in safe condition.

B. Guys
The general requirements for guys are covered in Rules 282 and 283.

C. Crossarm Strength

Crossarms shall be securely supported by bracing, if necessary, to withstand all expected loads to which they may be subjected, including line personnel working on them.

NOTE: Double crossarms are generally used at crossings, unbalanced corners, and dead ends, in order to permit conductor fastenings at two insulators to prevent slipping, although single crossarms might provide sufficient strength. To secure extra strength, double crossarms are frequently used, and crossarm guys are sometimes used.

D. Supply-Line Conductors

1. Size

Supply-line conductors shall be not smaller than the sizes listed in Table 263-1.

RECOMMENDATION: It is recommended that these minimum sizes for copper and steel be not used in spans longer than 150 feet for the heavy-loading district, and 175 feet for the medium- and light-loading districts.

Table 263-1. Grade N Minimum Sizes for Supply Line Conductors
(AWG for Copper and Aluminum; Stl. WG for Steel)

	Urban	Rural
Soft copper	6	8
Medium or hard-drawn copper	8	8
Steel	9	9
	Spans 150 ft or less	**Spans exceeding 150 ft**
Stranded aluminum:		
EC	4	2
ACSR	6	4
ALLOY	4	4
ACAR	4	2

E. Service Drops

1. Size of Open-Wire Service Drops

    a. Not over 750 volts. Service drops shall be as required by (1) or (2):

Table 263-2. Minimum Sizes of Service Drops Carrying 750 V or Less
(Voltages of trolley-contact conductors are voltage to ground. AWG used for aluminum copper wires; Stl. WG used for steel wire)

Situation	Copper wire		Steel wire	EC aluminum wire ②
	Soft drawn	Medium or hard drawn		
Alone	10	12	12	4
Concerned with communication conductor	10	12	12	4
Over supply conductors of				
0 to 750 V	10	12	12	4
750 V to 8.7kV①	8	10	12	4
Exceeding 8.7kV①	6	8	9	4
Over trolley-contact conductors				
0 to 750 V ac or dc	8	10	12	4
Exceeding 750 V ac or dc	6	8	9	4

① Installation of service drops of not more than 750 V above supply lines of more than 750 V should be avoided where practical.
② Where ACSR or aluminum alloy is used, the minimum size shall be No. 6 wire.

      (1) Spans not exceeding 150 feet. Sizes shall not be smaller than those specified in Table 263-2.

      (2) Spans exceeding 150 feet. Sizes shall not be smaller than required for grade C (Rule 261H1).

    b. **Exceeding 750 volts.** Sizes of service drops of more than 750 volts shall not be less than required for supply-line conductors of the same voltage.

  2. **Tension of Open-Wire Service Drops**

The tension of the service drop conductors shall not exceed the strength of the conductor attachment or its support under the expected loadings.

  3. **Cabled Service Drops**

Service conductors may be grouped together in a cable, provided the following requirements are met:

    a. **Size**

The size of each conductor shall not be less than required for drops of separate conductors (Rule 263E1).

    b. **Tension of Cabled Service Drops**

The tension of the service drop conductors shall not exceed the strength of the conductor attachment or its support under the expected loadings.

F. **Trolley-Contact Conductors**

In order to provide for wear, no trolley-contact conductors shall be installed of less size than No. 0 AWG, if of copper, or No. 4 AWG, if of silicon bronze.

G. **Communication Conductors**

There are no specific requirements for grade N communication line conductors or service drops.

Section 27. Line Insulation

270. Application of Rule

These requirements apply only to open conductor supply lines.

NOTE 1: See Rule 243C6.
NOTE 2: See Rule 242E for insulation requirements for neutral conductors.

271. Material and Marking

Insulators for operation of supply circuits shall be made of wet process porcelain or other material which will provide equivalent or better electrical and mechanical performance. Insulators for use at or above 2.3 kilovolts between conductors shall be marked by the maker with his name or trademark and an identification mark or markings which will permit determination of the electrical and mechanical properties. The marking shall be applied so as not to reduce the electrical or mechanical strength of the insulator.

NOTE: The identifying marking can be either a catalog number, trade number, or any other means so that properties of the unit can be determined either through catalogs or other literature.

272. Ratio of Flashover to Puncture Voltage

Insulators shall be designed so that the ratio of their rated low frequency dry flashover voltage to low frequency puncture voltage is in conformance with applicable American National Standards. When a standard does not exist, this ratio shall not exceed 75 percent.

The applicable American National Standards are:
ANSI C29.1-1976 Test Methods for Electrical Power Insulators
ANSI C29.2-1977 Wet Process Porcelain and Toughened Glass Insulators (Suspension Type)
ANSI C29.3-1977, Wet Process Porcelain Insulators (Spool Type)
ANSI C29.4-1977, Wet Process Porcelain Insulators (Strain Type)
ANSI C29.5-1977, Wet Process Porcelain Insulators, Low- and Medium Voltage Pin Type
ANSI C29.6-1977 and C29.6a-1974, Wet Process Porcelain Insulators, High-Voltage Pin Type
ANSI C29.7-1977, Wet Process Porcelain Insulators, High Voltage Line-Post Type

EXCEPTION: Insulators specifically designed for use in areas of high atmospheric contamination may have a rated low frequency dry flashover voltage not more than 80 percent of their low frequency puncture voltage.

273. Insulation Level

The rated dry flashover voltage of the insulator or insulators, when tested in accordance with ANSI C29.1-1976 shall not be less than that shown in Table 273-1, unless based on a qualified engineering study. Higher insulation levels than those shown in Table 273-1, or other effective means,

Table 273-1. Insulation Level Requirements

Nominal voltage (between phases) (kV)	Minimum rated dry flashover voltage of insulators ① (kV)	Nominal voltage (between phases) (kV)	Minimum rated dry flashover voltage of insulators ① (kV)
0.75	5	46	125
2.4	20	69	175
6.9	39	115	315
13.2	55	138	390
23.0	75	161	445
34.5	100	230	640

① Interpolate for intermediate values.

shall be used where severe lightning, high atmospheric contamination, or other unfavorable conditions exist. Insulation levels for system voltages in excess of those shown shall be based on a qualified engineering study.

274. Factory Tests

Each insulator or insulating part thereof for use on circuits operating at or above 2.3 kilovolts between conductors shall be tested by the manufacturer in accordance with applicable American National Standards or, where such standards do not exist, other good engineering practices to assure their performance.

The applicable American National Standards are listed in Rule 272.

275. Special Insulator Applications

A. Insulators for Constant-Current Circuits

Insulators for use on constant-current circuits shall be selected on the basis of the rated full load voltage of the supply transformer.

B. Insulators for Single-Phase Circuits Directly Connected to Three-Phase Circuits

Insulators used on single-phase circuits directly connected to three-phase circuits (without intervening isolating transformers) shall have an insulation level not less than that required for the three-phase circuit.

276. Protection Against Arcing and Other Damage

In installing and maintaining insulators and conductors, precautions shall be taken to prevent as far as is practical any damage which might render the conductors or insulators liable to fall. Precautions shall also be taken to prevent, as far as is practical, any arc from forming or prevent any arc which might be formed from injuring or burning any parts of the supporting structures, insulators, or conductors.

277. Mechanical Strength of Insulators

Insulators shall withstand all the loads specified in Section 25 except those of Rule 250C without exceeding the following percentage of their rated ultimate strength:

Cantilever   40 percent
Compression 50 percent
Tension   50 percent

NOTE 1: The rated ultimate mechanical strength of suspension type insulators is considered to be the rated "combined mechanical and electrical strength."
NOTE 2: See ANSI C29.1-1961 (R1974) or the latest revision thereof.

278. Aerial Cable Systems

A. Electrical Requirements

1. Covered or insulated conductors not meeting the requirements of Rule 230C1, 230C2, or 230C3 shall be considered as bare conductors for all insulation requirements.
2. The insulators or insulating supports shall meet the requirements of Rule 273.
3. The systems shall be so designed and installed to minimize long term deterioration from electrical stress.

B. Mechanical Requirements

1. Insulators other than spacers used to support aerial cable systems shall meet the requirements of Rule 277.
2. Insulating spacers used in spacer cable systems shall withstand the loads specified in Section 25 (except those of Rule 250C) without exceeding 50 percent of their rated ultimate strength.

Section 28. Miscellaneous Requirements

280. Structures for Overhead Lines
  A. Supporting Structures
    1. Protection of Structures
      a. Mechanical Injury
        Appropriate physical protection shall be provided for supporting structures subject to vehicular traffic abrasion which would materially affect their strength.
      b. Climbing
        Readily climbable supporting structures, such as closely latticed poles or towers, including those attached to bridges, carrying open supply conductors energized at more than 300 volts, which are adjacent to roads, regularly travelled pedestrian thoroughfares, or places where persons frequently gather (such as schools or public playgrounds) shall be equipped with barriers to inhibit climbing by unqualified persons or posted with appropriate warning signs.

        EXCEPTION: This rule does not apply where the right-of-way is fenced.

      c. Fire
        Supporting structures shall be placed and maintained so as to be exposed as little as is practical to brush, grass, rubbish, or building fires.
      d. Attached to Bridges
        Supporting structures attached to bridges for the purpose of carrying open supply conductors exceeding 600 volts shall be posted with appropriate warning signs.
    2. Steps
    Steps permanently installed on supporting structures shall not be closer than 8 feet from the ground or other accessible surface.

    EXCEPTION: This rule does not apply where supporting structures are isolated.

    3. Identification
    Supporting structures, including those on bridges, on

which supply or communication conductors are maintained shall be so constructed, located, marked, or numbered so as to facilitate identification by employees authorized to work thereon. Date of installation of such structures should be recorded where practical by the owner.

4. Obstructions

Signs, posters, notices, and other attachments shall not be placed on supporting structures without concurrence of the owner. Supporting structures should be kept free from other climbing hazards such as tacks, nails, vines, and through bolts not properly trimmed.

5. Decorative Lighting

Attachment of decorative lighting on structures shall not be made without the concurrence of the owners and occupants.

B. Unusual Conductor Supports

Where conductors are attached to structures other than those used solely or principally for their support, all rules shall be complied with as far as they apply. Such additional precautions as may be deemed necessary by the administrative authority shall be taken to avoid damage to the structures or injury to the persons using them. The supporting of conductors on trees and roofs should be avoided.

281. Tree Trimming

A. General

1. Trees which may interfere with supply conductors should be trimmed or removed.

   NOTE: Normal tree growth, the combined movement of trees and conductors under adverse weather conditions, voltage, and sagging of conductors at elevated temperatures are among the factors to be considered in determining the extent of trimming required.

2. Where trimming or removal is not practical, the conductor should be separated from the tree with suitable materials or devices to avoid conductor damage by abrasion and grounding of the circuit through the tree.

B. At Line Crossings, Railroad Crossings, and Limited Access Highway Crossings

The crossing span and the adjoining span on each side of the crossing should be kept free from overhanging or decayed trees or limbs which otherwise might fall into the line.

282. Guying and Bracing

A. Where Used

When the loads to be imposed on supporting structures are greater than can be safely supported by the structures alone, additional strength shall be provided by the use of guys, braces, or other suitable construction. Such measures shall also be used where necessary to prevent undue increase of sags in adjacent spans as well as to provide sufficient strength for those supports on which the loads are considerably unbalanced, for example, at corners, angles, dead ends, large differences in span lengths, and changes of grade of construction.

B. Strength

The strength of the guy or brace shall meet the requirements of Section 26 for the applicable grade of construction.

C. Point of Attachment

The guy or brace should be attached to the structure as near as is practical to the center of the conductor load to be sustained. However, on lines exceeding 8.7 kilovolts the location of the guy or brace may be adjusted to minimize the reduction of the insulation offered by nonmetallic support arms and supporting structures.

D. Guy Fastenings

Guys having an ultimate strength of 2000 pounds or more and subject to small radius bends should be stranded and should be protected by suitable guy thimbles or their equivalent. Cedar and other softwood poles around which any guy having an ultimate strength of 10 000 pounds or more is wrapped should be protected by the use of suitable guy shims.

Where there is a tendency for the guy to slip off the shim, guy hooks or other suitable means of preventing this action should be used. Shims are not necessary in the case of supplementary guys, such as storm guys.

E. Guy Markers (Guy Guards)

The ground end of anchor guys, exposed to pedestrian traffic, shall be provided with a substantial and conspicuous marker not less than 8 feet long.

NOTE: Visibility of markers can be improved by the use of color or color patterns which provide contrast with the surroundings.

F. Electrolysis

Where anchors and rods are subject to electrolysis, suitable measures should be taken to minimize corrosion from this source.

G. **Anchor Rods**

1. Anchor rods shall be installed so as to be in line with the pull of the attached guy when under load.

   EXCEPTION: This is not required for anchor rods installed in rock or concrete.

2. The anchor rod assembly shall have an ultimate strength not less than that required of the guy.

283. Insulators in Guys Attached to Supporting Structures

A. Properties of Guy Insulators

1. Material

   Insulators shall be made of wet process porcelain, wood, glass fiber reinforced plastic or other material of suitable mechanical and electrical properties.

2. Electrical Strength

   The guy insulator shall have a rated dry flashover voltage at least double the nominal line voltage and a rated wet flashover voltage at least as high as the nominal line voltage between conductors of the guyed circuit. A guy insulator may consist of one or more units.

3. Mechanical Strength

   The rated ultimate strength of the guy insulator shall be at least equal to the rated breaking strength of the guy in which it is installed.

B. Use of Guy Insulators

1. Ungrounded guys attached to supporting structures carrying open supply conductors of more than 300 volts, or if exposed to such conductors, shall be insulated.

   NOTE: Guys grounded in accordance with Rule 215C2 need not be insulated.

   EXCEPTION: A guy insulator is not required if the guy is attached to a supporting structure on private right-of-way if all the supply circuits exceeding 300 volts meet the requirements of Rule 220B2.

2. Insulators shall be installed as follows:

   a. All insulators shall be located at least 8 feet above the ground.

   b. Where hazard would exist with one insulator, two or more guy insulators shall be placed so as to include, in so far as is practical, the exposed section of the guy between them.

    c. Insulators shall be so placed that in case any guy sags down upon another, the insulators will not become ineffective.

C. Corrosion Protection

An insulator in the guy strand used exclusively for the elimination of corrosion of metal in ground rods, anchors, anchor rods, or pipe in an effectively grounded system, shall not be classified as a guy insulator and shall not reduce the mechanical strength of the guy.

284. Span-Wire Insulators

A. Properties of Span-Wire Insulators

1. Material

Insulators shall be made of wet process procelain, wood, fiberglass, or other material of suitable mechanical and electrical properties.

2. Insulation Level

The insulation level of span-wire insulators shall meet the requirements of Rule 274.

A hanger insulator, where used to provide single insulation as permitted by Rule 284B, shall meet the requirements of Rule 274.

3. Mechanical Strength

The rated ultimate strength of the span-wire insulator shall be at least equal to the rated breaking strength of the span wire in which it is installed.

B. Use of Span-Wire Insulators

1. All span wires, including bracket span wires, shall have a suitable insulator (in addition to an insulated hanger if used) inserted between each point of support of the span wire and the luminaire or trolley-contact conductor supported.

EXCEPTION 1: Single insulation, as provided by an insulated hanger, may be permitted when the span wire or bracket is supported on wood poles supporting only trolley, railway feeder, or communication conductors used in the operation of the railway concerned.

EXCEPTION 2: Insulators are not required if the span wire is effectively grounded.

EXCEPTION 3: This rule does not apply to insulated feeder taps used as span wires.

2. In case insulated hangers are not used, the insulator shall be located so that in the event of a broken wire the energized part of the span wire cannot be reached from the ground.

285. Overhead Conductors

A. Identification

All conductors of electric-supply and communication lines should, as far as is practical, be arranged to occupy uniform positions throughout, or shall be constructed, located, marked, numbered, or attached to distinctive insulators or crossarms, so as to facilitate identification by employees authorized to work thereon. This does not prohibit systematic transposition of conductors.

B. Branch Connections

1. Connections to circuits, service loops, and equipment in overhead construction shall be accessible to authorized employees.

2. Connections shall be supported and placed so that swinging or sagging cannot bring them in contact with other conductors or interfere with the safe use of pole steps, or reduce the climbing or lateral working space.

286. Equipment on Supporting Structures

A. Identification

All equipment of electric-supply and communication lines should be arranged to occupy uniform positions throughout or shall be constructed, located, marked, or numbered so as to facilitate identification by employees authorized to work thereon.

B. Location

All supply and communication equipment such as transformers, regulators, capacitors, amplifiers, loading coils, surge arresters, switches, etc, when located below conductors or other attachments, shall be mounted outside of the climbing space required in Section 23.

C. Guarding

Exposed energized parts of equipment such as switches, circuit breakers, surge arresters, etc, shall be enclosed or guarded if all of the following conditions apply:

1. The equipment is located below the top conductor support.

2. The equipment is located on the climbing side of the structure.

3. The requirements of Rule 422B, Part 4, of this code, cannot be met.

D. Working Clearance

All parts of equipment such as switches, fuses, transformers, surge arresters, etc, or other connections which may require operation or adjustment while energized and exposed at such times, shall be arranged so that in adjustment or operation no portion of the body, including the hands, need be brought closer to any exposed energized parts or conductors than permitted in Rules 422B or 427C, part 4 of this code.

E. Clearance Above Ground

Equipment shall be mounted at not less than the following heights above ground, measured to the lower projection of such equipment:

1. Equipment cases which are effectively grounded, or ungrounded cases which contain equipment connected to circuits of not more than 150 volts:

    Over traveled portions of roadway   16 feet
    Over shoulder of roadway   15 feet
    Over walkways   10 feet

    EXCEPTION 1: The bottom of the housing of traffic control signals suspended over the traveled portion of the roadway shall be not less than 15 feet nor more than 19 feet above the grade at the center of the roadway.
    EXCEPTION 2: Effectively grounded equipment cases such as fire alarm boxes, traffic control boxes, or meters may be mounted over a walkway at a lower level for accessibility provided such equipment does not unduly obstruct the walkway.

2. Ungrounded equipment cases which contain equipment connected to circuits of more than 150 volts shall have the same clearances above ground as specified for rigid live parts in Rule 232C.

F. Clearances from Buildings, Bridges, or Other Structures

1. Effectively grounded equipment cases may be located on or adjacent to buildings, bridges, or other structures provided that all exposed live parts of such equipment are located so that the clearances for open supply line conductors as specified in Rules 234C, 234D, and 234F are maintained.

2. Equipment cases which are not effectively grounded shall be located so that the clearances for open supply line conductors of Rules 234C, 234D, and 234F are maintained.

3. Equipment cases shall be located so as not to serve as a means of approach to exposed live parts by unqualified persons.

G. Street and Area Lighting

1. All exposed ungrounded conductive parts of luminaires and their supports which are not insulated from current-carrying parts shall be maintained at not less than 20 inches from the surface of their supporting structure:

   EXCEPTION 1: This may be reduced to 5 inches if located on the side of the structure opposite the designated climbing space.

   EXCEPTION 2: This does not apply where the equipment is located at the top or other vertical portion of the structure which is not subject to climbing.

2. The lowering rope or chain for luminaires arranged to be lowered for examination or maintenance shall be of a material and strength designed to withstand climatic conditions and to sustain the luminaire safely. The lowering rope or chain, its supports, and fastenings shall be examined periodically.

3. Insulators, as specified in Rule 283A, should be inserted at least 8 feet from the ground in metallic suspension ropes or chains supporting lighting units of series circuits.

4. A suitable device shall be provided by which each lamp on series lighting circuits of more than 300 volts may be safely disconnected from the circuit before the lamp is handled.

   EXCEPTION: This rule does not apply where the lamps are always worked on from suitable insulated platforms or aerial lift devices, or handled with suitable insulated tools, and treated as under full voltage of the circuit concerned.

287. Communications Protective Requirements

A. Where Required

Where communications apparatus is handled by other than qualified persons, it shall be protected by one or more of the means listed in part B of this rule if such apparatus is permanently connected to lines subject to any of the following:

1. Lightning.
2. Possible contact with supply conductors whose voltage to ground exceeds 300 volts.
3. Transient rise in ground potential exceeding 300 volts.
4. Steady state induced voltage of a hazardous level.

NOTE: When communications cables will be in the vicinity of supply stations where large ground currents may flow, the effect of these currents on communications circuits should be evaluated.

B. Means of Protection

Where communications apparatus is required to be protected under Part A of this rule, protective means adequate to withstand the voltage expected to be impressed shall be provided by insulation, protected where necessary by arresters used in conjunction with fusible elements. Severe conditions may require the use of additional devices such as auxiliary arresters, drainage coils, neutralizing transformers, or isolating devices.

288. **Circuits of One Class Used Exclusively in the Operation of Circuits of Another Class**

A. Overhead Communication Circuits Used Exclusively in the Operation of Supply Circuits

    1. Communication circuits used exclusively in the operation of supply lines may be run either as ordinary communication circuits or as supply circuits under the conditions specified in provisions 3 and 4 of this rule, respectively. After the selection of the type of communication circuit construction and protection for a section, such construction and protection shall be consistently adhered to throughout the extent of such section of the communication system.

    2. Communication circuits used in operation of supply lines shall be isolated or guarded at all points so as to be inaccessible to the public.

    3. Communication circuits used in the operation of supply lines may be run as ordinary communication conductors under the following conditions:

        a. Where such circuits are below supply conductors in the operation of which they are used (including high-voltage trolley feeders) at crossings, conflicts, or on commonly used poles, provided:

        (1) Such communication circuits occupy a position below all other supply conductors or equipment at crossings, conflicts, or on commonly used poles.

        (2) Such communication circuits and their connected equipment are adequately guarded and are accessible only to authorized persons.

b. Where such circuits are below supply conductors in the operation of which they are used and are above other supply or communication conductors at wire crossings, conflicts, or on the same poles, provided the communication circuits are protected by fuseless surge arresters, drainage coils, or other suitable devices to prevent the communication circuit voltage from normally exceeding 400 volts to ground.

NOTE: The grades of construction for communication conductors with inverted levels apply.

4. Communication circuits used in the operation of supply lines shall comply with all requirements for the supply lines with which they are used, where they do not comply with the provisions of 3a or 3b above.

EXCEPTION 1: If the voltage of the supply conductors concerned exceeds 8.7 kilovolts, the communication conductors need only meet the requirements for supply conductors of 5 to 8.7 kilovolts.

EXCEPTION 2: Where the supply conductors are required to meet grade C, the size of the communication conductors may be the same as for grade D (see Rule 26212) for spans up to 150 feet.

B. Supply Circuits Used Exclusively in the Operation of Communication Circuits

Circuits used for supplying power solely to apparatus forming part of a communications system shall be installed as follows:

1. Open wire circuits shall have the grades of construction, clearances, insulation, etc, prescribed elsewhere in these rules for supply or communication circuits of the voltage concerned.

2. Special circuits operating at voltages in excess of 400 volts to ground and used for supplying power solely to communications equipment may be included in communications cables under the following conditions:

a. Such cables shall have a conductive sheath or shield which is effectively grounded and each such circuit shall be carried on conductors which are individually enclosed with an effectively grounded shield.

b. All circuits in such cables shall be owned or operated by one party and shall be maintained only by qualified personnel.

    c. Supply circuits included in such cables shall be terminated at points accessible only to qualified personnel.

    d. Communications circuits brought out of such cables, if they do not terminate in a repeater station or terminal office, shall be protected or arranged so that in the event of failure within the cable, the voltage on the communication circuit will not exceed 400 volts to ground.

    e. Terminal apparatus for the power supply shall be so arranged that the live parts are inaccessible when such supply circuits are energized.

EXCEPTION: The requirements of Rule 288 do not apply to the supply circuits of 600 volts or less where the transmitted power does not exceed 5 kilowatts and the installation complies with Rule 220B2.

289. Electric Railway Construction

A. Trolley-Contact Conductor Fastenings

All overhead trolley-contact conductors shall be supported and arranged so that the breaking of a single contact conductor fastening will not allow the trolley conductor live span wire, or current-carrying connection to come within 10 feet (measured vertically) from the ground, or from any platform accessible to the general public.

Span-wire insulation for trolley-contact conductors shall comply with Rule 284.

B. High Voltage Contact Conductors

Trolley-contact conductors energized at more than 750 volts shall be suspended so as to minimize the possibility of a break, and in such a way that, if broken at one point, the conductor will not come within 12 feet (measured vertically) of the ground, or any platform accessible to the public.

C. Third Rails

Third rails shall be protected by adequate guards composed of wood or other suitable insulating material.

EXCEPTION: This rule does not apply where third rails are on fenced right-of-way.

D. Prevention of Loss of Contact at Railroad Crossings at Grade

At crossings at grade with other railroads or other electrified railway systems, contact conductors shall be arranged as set forth in specifications 1, 2, 3, 4, and 5 following, whichever apply:

1. Where the crossing span exceeds 100 feet, catenary construction shall be used for overhead trolley-contact conductors.

2. When pole trolleys, using either wheels or sliding shoes, are used:

    a. The trolley-contact conductor shall be provided with live trolley guards of suitable construction; or

    b. The trolley-contact conductor should be at a uniform height above its own track throughout the crossing span and the next adjoining spans. Where it is not practical to maintain a uniform height, the change in height shall be made in a gradual manner.

    EXCEPTION: Rule 289D2 does not apply where the crossing is protected by signals or interlocking.

3. When pantograph type collectors are used, the contact conductor and track through the crossing should be maintained in a condition where rocking of pantograph-equipped cars or locomotives will not de-wire the pantograph. If this cannot be done, auxiliary contact conductors shall be installed. Wire height shall conform with Rule 289D2b.

4. Where two electrified tracks cross:

    a. When the trolley-contact conductors are energized from different supply circuits, or from different phases of the same circuit, the trolley-conductor crossover shall be designed to insulate both conductors from each other. The design shall not permit either trolley collector to contact any conductor or part energized at a different voltage than at which it is designed to operate.

    b. Trolley-contact crossovers used to insulate trolley conductors of the same voltage but of different circuit sections shall be designed to prevent both sections being simultaneously contacted by the trolley collector.

5. When third rail construction is used, and the length of the third rail gap at the crossings is such that a car or locomotive stopping on the crossing can lose propulsion power, the crossing shall be protected by signals or interlocking.

E. Guards Under Bridges

Trolley guards of suitable construction shall be provided where the trolley-contact conductor is so located that a trolley pole leaving the conductor can make simultaneous contact between it and the bridge structure.

American National Standard

National Electrical Saftey Code

Part 3 (Sections 30-39). Safety Rules for the Installation and Maintenance of Underground Electric-Supply and Communication Lines

Secretariat
Institute of Electrical and Electronics Engineers

3

PART 3. Safety Rules for the Installation and Maintenance of Underground Electric-Supply and Communication Lines

Section 30. Purpose, Scope, and Application of Rules

300. Purpose

The purpose of Part 3 of this code is the practical safeguarding of persons during the installation, operation, or maintenance of underground or buried supply and communication cables and associated equipment.

301. Scope

Part 3 of this code covers supply and communication cables and equipment in underground or buried systems. The rules cover the associated structural arrangements and the extension of such systems into buildings. It also covers the cables and equipment employed primarily for the utilization of electric power when such cables and equipment are used by the utility in the exercise of its function as a utility. They do not cover installations in electric supply stations.

Section 31. General Requirements Applying to Underground Lines

310. Rule 310 not used in this edition.

311. Installation and Maintenance

A. Persons responsible for underground facilities shall be in a position to indicate the location of their facilities.

B. Reasonable advance notice should be given to owners or operators of other proximate facilities which may be adversely affected by new construction or changes in existing facilities.

312. Accessibility

All parts which must be examined or adjusted during operation shall be arranged so as to be readily accessible to authorized persons by the provision of adequate working spaces, working facilities, and clearances.

313. Inspection and Tests of Lines and Equipment

A. When in Service
  1. Initial Compliance with Safety Rules
     Lines and equipment shall comply with these safety rules upon being placed in service.
  2. Inspection
     Accessible lines and equipment shall be inspected from time to time by the responsible party at such intervals as experience has shown to be necessary.
  3. Tests
     When considered necessary, lines and equipment shall be subjected to practical tests to determine required maintenance.
  4. Record of Defects
     Any defects affecting compliance with this Code revealed by inspection, if not promptly corrected, shall be recorded; such record shall be maintained until the defects are corrected.
  5. Remedying Defects
     Lines and equipment with recorded defects which would endanger life or property, shall be properly repaired, disconnected, or isolated.

B. When Out of Service
  1. Lines Infrequently Used
     Lines and equipment infrequently used shall be inspected or tested as necessary before being placed into service.
  2. Lines Temporarily Out of Service
     Lines and equipment temporarily out of service shall be maintained in a safe condition.
  3. Lines Permanently Abandoned
     Lines and equipment permanently abandoned shall be removed or maintained in a safe condition.

314. Grounding of Circuits and Equipment

A. Methods
   The methods to be used for grounding of circuits and equipment are given in Section 9.

B. Conductive Parts to Be Grounded

Cable sheaths and shields (except conductor shields), equipment frames and cases (including pad-mounted devices), and lamp posts shall be effectively grounded. Ducts and riser guards of conductive material which are exposed to probable contact with conductors of more than 300 V to ground shall be effectively grounded.

EXCEPTION 1: This rule does not apply to parts which are 8 ft or more above readily accessible surfaces or are otherwise isolated or guarded.

EXCEPTION 2: This rule does not apply to ducts and riser guards which contain cables having effectively grounded sheaths or shields in contact with the duct or guard.

C. Use of Earth as Part of Circuit

Supply circuits shall not be designed to use the earth normally as the sole conductor for any part of the circuit.

315. Communication Protective Requirements

A. Where Required

Where communications apparatus is handled by other than qualified persons, it shall be protected by one or more of the means listed in Rule 315 B if such apparatus is permanently connected to lines subject to any of the following:

1. Lightning.
2. Possible contact with supply conductors whose voltage exceeds 300 V.
3. Transient rise in ground potential exceeding 300 V.
4. Steady-state induced voltage of a hazardous level.

NOTE: When communications cables will be in the vicinity of supply stations where large ground currents may flow, the effect of these currents on communications circuits should be evaluated.

B. Means of Protection

Where communications apparatus is required to be protected under Rule 315 A, protective means adequate to withstand the voltage expected to be impressed shall be provided by insulation, protected where necessary by arresters. Severe conditions may require the use of additional devices such as auxiliary arresters, drainage coils, neutralizing transformers, or isolating devices.

316. Induced Voltage

Rules covering electrical influence and susceptiveness have

not been detailed in this code. Steady-state induced hazards from proximate facilities shall be eliminated. Transient induced hazards should be minimized insofar as practical. Cooperative procedures are recommended.

Section 32. Underground Conduit Systems

NOTE: While it is often the practice to use *duct* and *conduit* interchangeably, *duct,* as used herein, is a single enclosed raceway for conductors or cable; *conduit* is a structure containing one or more ducts; and *conduit system* is the combination of conduit, conduits, manholes, hand-holes, and/or vaults joined to form an integrated whole.

320. Location
A. Routing
1. General
   a. Conduit systems should be subject to the least disturbance practical. Conduit systems extending parallel to other subsurface structures should not be located directly over or under other subsurface structures. If this is not practical, the rule on clearances, as stated in Rule 320 B, should be followed.
   b. Conduit alignment should be such that there are no protrusions which would be harmful to the cable.
   c. When bends are required, the minimum radius shall be sufficiently large to prevent damage to cable being installed in the conduit.

   RECOMMENDATION: The maximum change of direction in any plane between lengths of straight rigid conduit without the use of bends should be limited to 5°.

2. Natural Hazards
   Routes through unstable soils such as mud, shifting soil, etc, or through highly corrosive soils, should be avoided. If construction is required in these soils, the conduit should be constructed in such a manner as to minimize movement and or corrosion or both.

3. Highways and Streets

When conduit must be installed longitudinally under the roadway, it should be installed in the shoulder or, to the extent practical, within the limits of one lane of traffic.

4. Bridges and Tunnels

The conduit system shall be located so as to minimize the possibility of damage by traffic. It should be located to provide safe access for inspection or maintenance of both the structure and the conduit system.

5. Crossing Railroad Tracks

a. The top of the conduit system should be located not less than 36 in below the top of the rails of a street railway or 60 in below the top of the rails of a railroad. Where unusual conditions exist or where proposed construction would interfere with existing installations, a greater depth than specified above may be required.

EXCEPTION: Where this is impractical, or for other reasons, this clearance may be reduced by agreement between the parties concerned. In no case, however, shall the top of the conduit or any conduit protection extend higher than the bottom of the ballast section which is subject to working or cleaning.

b. At crossings under railroads, manholes, handholes, and vaults should not, where practical, be located in the roadbed.

6. Submarine Crossing

Submarine crossings should be routed, installed, or both so they will be protected from erosion by tidal action or currents. They should not be located where ships normally anchor.

B. Clearances from Other Underground Installations

1. General

The clearance between a conduit system and other underground structures paralleling it should be as large as necessary to permit maintenance of the system without damage to the paralleling structures. A conduit which crosses over another subsurface structure shall have a minimum clearance sufficient to prevent damage to either structure. These clearances should be determined by the parties involved.

EXCEPTION: When conduit crosses a manhole, vault, or subway tunnel roof, it may be supported directly on the roof with the concurrence of all parties involved.

2. Separations Between Supply and Communications Conduit Systems

Conduit systems to be occupied by communications conductors shall be separated from conduit systems to be used for supply systems by:

   a. 3 in of concrete.

   b. 4 in of masonry.

   c. 12 in of well tamped earth.

EXCEPTION: Lesser separations may be used where the parties concur.

3. Sewers, Sanitary and Storm

   a. If conditions require a conduit to be installed parallel to and directly over a sanitary or storm sewer, it may be done provided both parties are in agreement as to the method.

   b. Where a conduit run crosses a sewer it shall be designed to have suitable support on each side of the sewer to prevent transferring any direct load onto the sewer.

4. Water Lines

Conduit should be installed as far as is practical from a water main in order to protect it from being undermined if the main breaks. Conduit which crosses over a water main shall be designed to have suitable support on each side as required to prevent transferring any direct loads onto the main.

5. Fuel Lines

Conduit should have sufficient clearance from fuel lines to permit the use of pipe maintenance equipment. Conduit and fuel lines shall not enter the same manhole.

6. Steam Lines

Conduit should be so installed as to prevent detrimental heat transfer between the steam and conduit systems.

321. Excavation and Backfill

A. Trench

The bottom of the trench should be undisturbed, tamped, or relatively smooth earth. Where the excavation is in rock, the conduit should be laid on a protective layer of clean tamped backfill.

B. Quality of Backfill

All backfill should be free of materials that may damage the conduit system.

RECOMMENDATION: Backfill within 6 in of the conduit should be free of solid material greater than 4 in in maximum dimension or with sharp edges likely to damage it. The balance of backfill should be free of solid material greater than 8 in in maximum dimension. Backfill material should be adequately compacted.

322. Ducts and Joints

A. General

1. Duct material shall be corrosion resistant and suitable for the intended environment.

2. Duct materials, the construction of the conduit, or both shall be designed so that a cable fault in one duct would not damage the conduit to such an extent that it would cause damage to cables in adjacent ducts.

3. The conduit system shall be designed to withstand external forces to which it may be subjected by the surface loadings set forth in Rule 323 A except that impact loading may be reduced one third for each foot of cover so no impact loading need be considered when cover is 3 ft or more.

4. The internal finish of the duct shall be free of sharp edges or burrs which could damage supply cable.

B. Installation

1. Restraint

Conduit, including terminations and bends, should be suitably restrained by backfill, concrete envelope, anchors, or other means to maintain its design position under stress of installation procedures, cable pulling operations, and other conditions such as settling and hydraulic or frost uplift.

2. Joints

Ducts shall be joined in a manner sufficient to prevent solid matter from entering the conduit line. Joints shall form a sufficiently continuous smooth interior surface between joining duct sections so that supply cable will not be damage when pulled past the joint.

3. Externally Coated Pipe

When conditions are such that externally coated pipe is required, the coating shall be corrosion resistant and should be inspected, tested, or both, to see that the coating ˙ is continuous and intact prior to backfill. Precautions shall be taken to prevent damage to the coating when backfilling.

4. Building Walls

Conduit installed through a building wall shall have internal and external seals intended to prevent the entrance of gas into the building insofar as practical. The use of seals may be supplemented by gas venting devices in order to minimize building up of positive gas pressures in the conduit.

5. Bridges

a. Conduit installed in bridges shall include the capability to allow for expansion and contraction of the bridge.

b. Conduits passing through a bridge abutment should be installed so as to avoid or resist any shear due to soil settlement.

c. Conduit of conductive material installed on bridges shall be effectively grounded.

6. In Vicinity of Manholes

Conduit should be installed on compacted soil or otherwise supported when entering a manhole to prevent shear stress on the conduit at the point of manhole entrance.

323. Manholes, Handholes and Vaults

A. Strength

Manholes, handholes, and vaults shall be designed to sustain all expected loads which may be imposed upon the structure. The horizontal design loads, vertical design loads, or both shall consist of dead load, live load, equipment load, impact, load due to water table, frost, and any other load expected to be imposed upon the structure, to occur adjacent to the structure, or both. The structure shall sustain the combination of vertical and lateral loading that produces the maximum shear and bending moments in the structure.

1. In roadway areas, the live load shall consist of the weight of a moving tractor-semitrailer truck illustrated in Fig 1. The vehicle wheel load shall be considered applied to an area as indicated in Fig 2. In the case of multilane pavements, the structure shall sustain the combination of loadings which result in vertical and lateral structure loadings which produce the maximum shear and bending moments in the structure.

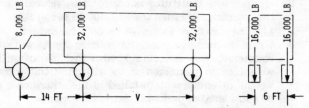

V = Variable spacing, 14 ft to 30 ft inclusive. Spacing to be used is that which results in vertical and lateral structure loading which produces the maximum shears and bending moments in the structure.

Fig 1
Roadway Vehicle Load

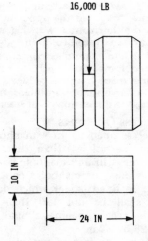

Fig 2
Wheel Load Area

NOTE: Loads imposed by equipment used in road construction may exceed loads to which the completed road may be subjected.

2. In designing structures not subject to vehicular loading, the minimum live load shall be 300 lb per sq ft.
3. Live loads shall be increased by 30% for impact.
4. When hydraulic, frost, or other uplift will be encountered, the structure shall either be of sufficient weight or so restrained as to withstand this force. The weight of equipment installed in the structure is not to be considered as part of the structure weight.
5. Where pulling iron facilities are furnished, they should be installed with a factor of safety of 2 based on the expected load to be applied to the pulling iron.

B. Dimensions

Manholes shall meet the following requirements:

A clear working space sufficient for performing the necessary work shall be maintained. The horizontal dimensions of the clear working space shall be not less than 3 ft. The vertical dimensions shall be not less than 6 ft except in manholes where the opening is within 1 ft, horizontally, of the adjacent interior side wall of the manhole.

EXCEPTION 1: Where one boundary of the working space is an unoccupied wall and the opposite boundary consists of cables only, the horizontal working space between these boundaries may be reduced to 30 in.

EXCEPTION 2: In manholes containing only communications cables, equipment, or both, one horizontal dimension of the working space may be reduced to not less than 2 ft provided the other horizontal dimension is increased so that the sum of the two dimensions is at least 6 ft.

C. Manhole Access Openings

1. Round access openings in a manhole containing supply cables shall be not less than 26 in in diameter. Round access openings in any manhole containing communication cables only, or manholes containing supply cables and having a fixed ladder which does not obstruct the opening, shall be not less than 24 in in diameter. Rectangular access openings should have dimensions not less than 26 by 22 in.
2. Openings shall be free of protrusions which will injure personnel or prevent quick egress.

D. **Covers**

   1. Manholes and handholes, when not being worked in, shall be securely closed by covers of sufficient weight or proper design so they cannot be easily removed without tools.

   2. Covers should be suitably designed or restrained so that they cannot fall into manholes or protrude into manholes sufficiently far to contact cable or equipment.

   3. Strength of covers and their supporting structure shall be at least sufficient to sustain the applicable loads of Rule 323 A.

E. **Access**

   1. Vault or manhole openings shall be located so that safe access can be provided. When in the highway, they should be located outside of the paved roadway when practical. They should be located outside the area of street intersections and crosswalks whenever practical to reduce the traffic hazards to the men working at these locations.

   2. a. Personnel access openings in vaults or manholes should be located so that they are not directly over the cable or equipment. Where these openings interfere with curbs, etc, they can be located over the cable if one of the following is provided:

      (1) A conspicuous warning sign.

      (2) A protective barrier over the cable.

      (3) A fixed ladder.

     b. In vaults, other types of openings may be located over equipment to facilitate work on this equipment.

F. **Access Doors**

   1. Where accessible to the public, access doors to utility tunnels and vaults shall be locked unless qualified persons are in attandance to prevent entry by unqualified persons.

   2. Such doors shall be designed so that a person on the inside may exit when the door is locked from the outside.

   EXCEPTION: This rule does not apply where the only means of locking is by padlock and the latching system is so arranged that the padlock can be closed on the latching system to prevent locking from the outside.

G. Ladder Requirements

Fixed ladders shall be corrosion resistant. For portable ladders, see Part 4 of this code.

RECOMMENDATION: See ANSI A14.1-1968, ANSI A14.2-1972, and ANSI A14.3-1956 on ladders.

H. Drainage

Where drainage is into sewers, suitable traps or other means should be provided to prevent entrance of sewer gas into manholes, vaults, or tunnels.

I. Ventilation

Adequate ventilation to open air shall be provided for manholes, vaults, and tunnels, having an opening into enclosed areas used by the public. Where such enclosures house transformers, switches, regulators, etc, the ventilating system shall be cleaned at necessary intervals.

EXCEPTION: This does not apply to enclosed areas under water or in other locations where it is impractical to comply.

J. Mechanical Protection

Supply cables and equipment should be installed or guarded in such a manner as to avoid damage by objects falling or being pushed through the grating.

K. Identification

Manhole and handhole covers should have an identifying mark which will indicate ownership or type of utility.

Section 33. Supply Cable

330. General

RECOMMENDATION: Cable should be capable of withstanding tests applied in accordance with an applicable standard issued by a recognized organization such as the American National Standard Institute, Association of Edison Illuminating Companies, the Insulated Cable Engineers Association, the National Electrical Manufacturers Association, or the American Society for Testing and Materials.

A.

The design and construction of conductors, insulation, sheath, jacket, and shielding shall include consideration of mechanical, thermal, environmental, and electrical stresses which are expected during installation and operation.

B. Cable shall be designed and manufactured to retain specified dimensions and structural integrity during manufacture, reeling, storage, handling, and installation.

C. Cable shall be designed and constructed in such a manner that each component is protected from harmful effects of other components.

D. The conductor, insulation, and shielding shall be designed to withstand the effects of the expected magnitude and duration of fault current, except in the immediate vicinity of the fault.

331. Sheaths and Jackets

Sheaths, jackets, or both shall be provided when necessary to protect the insulation or shielding from moisture or other adverse environmental conditions.

332. Shielding

A. General

1. Conductor shielding should, and insulation shielding shall, be provided as specified by an applicable document issued by a nationally recognized cable standardization organization.

   NOTE: Typical cable standardization organizations include: The Association of Edison Illuminating Companies, The Insulated Cable Engineers Association and The National Electrical Manufacturers Association.

   EXCEPTION: Shielding is not required for short jumpers which do not contact a grounded surface within enclosures or vaults, provided the jumpers are guarded or isolated.

2. Insulation shielding may be sectionalized provided that each section is effectively grounded.

B. Material

1. The shielding system may consist of semiconducting materials, nonmagnetic metal, or both. The shielding adjacent to the insulation shall be designed to remain in intimate contact with the insulation under all operating conditions.

2. Shielding material shall either be designed to resist excessive corrosion under the expected operating conditions or shall be protected.

333. Cable Accessories and Joints

A. Cable accessories and joints shall be designed to withstand the mechanical, thermal, environmental, and electrical stresses expected during operation.

B. Cable accessories and joints shall be designed and constructed in such a manner that each component of the cable and joint is protected from harmful effects of the other components.

C. Cable accessories and joints shall be designed and constructed to maintain the structural integrity of the cables to which they are applied and to withstand the magnitude and duration of the fault current expected during operation, except in the immediate vicinity of the fault.

D. For insulating joints, see Rule 332A2.

Section 34. Cable in Underground Structures

340. General

A. Section 33 shall apply to supply cable in underground structures.

B. On systems operating above 2 kV to ground, the design of the conductors or cables installed in nonmetallic conduit should consider the need for an effectively grounded shield, a sheath, or both.

341. Installation

A. General
1. Bending of the supply cable during handling, installation, and operation shall be controlled to avoid damage.
2. Pulling tensions and sidewall pressures of the supply cable should be limited to avoid damage.

   NOTE: Manufacturers' recommendations may be used as a guide.
3. Ducts should be cleaned of foreign material which could damage the supply cable during pulling operations.
4. Cable lubricants shall not be detrimental to cable or conduit systems.
5. On slopes or vertical runs, consideration should be given to restraining cables to prevent downhill movement.

6. Supply, control, and communication cables shall not be installed in the same duct unless the cables are maintained or operated by the same utility.

B. Cable in Manholes and Vaults
   1. Supports
      a. Cable supports shall be designed to withstand both live and static loading and should be compatible with the environment.
      b. Supports shall be provided to maintain specified separation between cables.
      c. Horizontal runs of supply cables shall be supported at least 3 in above the floor, or be suitably protected.

         EXCEPTION: This rule does not apply to grounding or bonding conductors.

      d. The installation should allow cable movement without destructive concentration of stresses. The cable should remain on supports during operation.

         NOTE: Special protection may be necessary at the duct entrance.

   2. Separation
      a. Adequate working space shall be provided in accordance with Rule 323 B.
      b. Between supply and communication facilities (cable, equipment, or both).
         (1) Where cable, equipment, or both are to be installed in a joint-use manhole or vault, it shall be done only with the concurrence of all parties concerned.
         (2) Supply and communication cables should be racked from separate walls. Crossings should be avoided.
         (3) Where supply and communication cables must be racked from the same wall, the supply cables should be racked below the communication cables.
         (4) Supply and communication facilities shall be installed to permit access to either without moving the other.
         (5) Clearances shall be maintained as specified in Table 1.

   3. Identification
      a. General
         (1) Cables shall be permanently identified by tags or

Table 1.
Minimum Separation Between Supply and Communications
Facilities in Joint-Use Manholes and Vaults

Phase-to-Phase Supply Voltage	Inches (Surface to Surface)
0 to 15 000	6
15 001 to 50 000	9
50 001 to 120 000	12
120 001 and above	24

EXCEPTION 1: These separations do not apply to grounding conductors.

EXCEPTION 2: These separations may be reduced by mutual agreement between the parties concerned when suitable barriers or guards are installed.

  otherwise at each manhole or other access opening of the conduit system.

  EXCEPTION: This requirement does not apply where the position of a cable, in conjunction with diagrams or maps supplied to workmen, gives sufficient identification.

 (2) All identification shall be of a corrosion-resistant material suitable for the environment.

 (3) All identification shall be of such quality and located so as to be readable with auxiliary lighting.

b. Joint-Use Manholes

 (1) Where cables in a manhole are maintained or operated by different utilities or are of supply and communication usage, they shall be permanently marked as to company, type of use, or both.

342. Grounding and Bonding

A. Insulation shielding of cable and joints shall be effectively grounded.

B. Cable sheaths or shields which are connected to ground at a manhole shall be bonded or connected to a common ground.

C. Bonding and grounding leads shall be of a corrosion resistant material suitable for the environment or suitably protected.

343. Fireproofing

A. Although fireproofing is not a requirement, it may be provided in accordance with each utility's normal service reliability practice to provide protection fron external fire.

344. Communication Cables Containing Special Supply Circuits

A. Special circuits operation at voltages in excess of 400 V to ground and used for supplying power solely to communications equipment may be included in communications cables under the following conditions:

1. Such cables shall have a conductive sheath or shield which shall be effectively grounded and each such circuit shall be carried on conductors which are individually enclosed with an effectively grounded shield.

2. All circuits in such cables shall be owned or operated by one party and shall be maintained only by qualified personnel.

3. Supply circuits included in such cables shall be terminated at points accessible only to qualified employees.

4. Communications circuits brought out of such cables, if they do not terminate in a repeater station or terminal office, shall be protected or arranged so that in event of a failure within the cable, the voltage on the communications circuit will not exceed 400 V to ground.

5. Terminal apparatus for the power supply shall be so arranged that live parts are inaccessible. when such supply circuits are energized.

6. Such cables shall be identified, and the identification shall meet the pertinent requirements of Rule 341 B 3.

EXCEPTION: The requirements of Rule 344 A do not apply to supply circuits of 550 V or less which carry power not in excess of 3200 W.

Section 35. Direct Buried Cable

350. General

A. Section 33 shall apply to direct buried supply cable.

B. Cables operating above 600 V to ground shall have a continuous shield, sheath, or concentric neutral which is effectively grounded.

C. Cables of the same circuit operating below 600 V to ground and without an effectively grounded shield or sheath shall be placed in close proximity (no intentional separation) to each other.

D. Communications cables containing special circuits supplying power solely to communications equipment shall comply with the requirements of Rules 344 A 1 through 344 A 5.

351. Location and Routing

A. General
1. Cables should be located so as to be subject to the least disturbance practical. Cables to be installed parallel to other subsurface structures should not be located directly over or under other subsurface structure, but if this is not practical, the rules on clearances in Rule 352 should be followed.
2. Cables should be installed in as straight and direct a line as practical. Where bends are required, the minimum radius shall be sufficiently large to prevent damage to the cable being installed.
3. Cable systems should be routed so as to allow safe access for construction, inspection, and maintenance.
4. The location of structures in the path of the projected cable route shall, as far as practical, be determined prior to trenching, plowing, or boring operation.

B. Natural Hazards
Routes through unstable soil such as mud, shifting soils, corrosive soils, or other natural hazards, should be avoided. If burying is required through areas with natural hazards, the cables shall be constructed and installed in such a manner as to protect them from damage. Such protective measures should be compatible with other installations in the area.

C. Other Conditions
  1. Swimming Pools
      Supply cable should not be installed within 5 ft of a
      swimming pool or its auxiliary equipment.
  2. Buildings and Other Structures
      Cable should not be installed directly under building or
      storage tank foundations. Where a cable must be installed
      under such a structure, the structure shall be suitably
      supported to prevent transfer of a harmful load onto the
      cable.
  3. Railroad Tracks
      a. The installation of cable longitudinally under the
          ballast section for railroad tracks should be avoided.
          Where cable must be installed longitudinally under
          the ballast section of a railroad, it should be located
          at a depth of not less than 50 inches below the top of
          the rail.

          EXCEPTION: Where this is impractical, or for other
          reasons, this clearance may be reduced by agreement
          between the parties concerned.

          NOTE: Where unusual conditions exist or where
          proposed construction would interfere with existing
          installations, a greater depth than specified above
          would be required.

      b. Where a cable crosses under railroad tracks, the same
          clearances indicated in Rule 320A5 shall apply.
  4. Highways and Streets
      The installation of cable longitudinally under traveled
      surfaces of highways and streets should be avoided. When
      cable must be installed longitudinally under the roadway,
      it should be installed in the shoulder or, if this is not
      practical, within the limits of one lane of traffic to the
      extent practical.
  5. Submarine Crossings
      Submarine crossings should be routed, installed, or both,
      so they will be protected from erosion by tidal action or
      currents. They should not be located where ships
      normally anchor.

352. Clearances From Other Underground Structures (sewers, water lines, fuel lines, building foundations, steam lines, other supply or communication conductors not in random separation, etc)

A. Horizontal Clearance

The horizontal clearance between direct buried cable and other underground structures shall be controlled at a minimum of 12 inches or larger as necessary to permit access to and maintenance of either facility without damage to the other. Installations with less than 12 inches horizontal separation shall conform with requirements of Rule 352 C, Rule 354, or both.

B. Crossings

1. Where a cable crosses under another underground structure, the structure shall be suitably supported to prevent transfer of a harmful load onto the cable system.
2. Where a cable crosses over another underground structure, the cable shall be suitably supported to prevent transfer of a harmful load onto the structure.
3. Adequate support may be provided by installing the facilities with sufficient vertical separation.
4. Adequate vertical clearance shall be maintained to permit access to and maintenance of either facility without damage to the other. A vertical clearance of 12 in is, in general, considered adequate but the parties involved may agree to a lesser separation.

C. Parallel Facilities

If conditions require a cable system to be installed with less than 12 in horizontal separation or directly over and parallel to another underground structure (or another underground structure installed directly over and parallel to a cable), it may be done providing all parties are in agreement as to the method. Adequate vertical clearance shall be maintained to permit access to and maintenance of either facility without damage to the other.

D. Thermal Protection

Cable should be installed with sufficient clearance from other underground structures, such as steam or cryogenic lines, to avoid thermal damage to the cable. Where it is not practical to provide adequate clearance, a suitable thermal barrier shall be placed between the two facilities.

353. Installation

A. Trenching

The bottom of the trench receiving direct buried cable should be relatively smooth undisturbed earth, well tamped earth, or sand. When excavation is in rock or rocky soils, the cable should be laid on a protective layer of well tamped backfill. Backfill within 4 inches of the cable should be free of materials that may damage the cable. Backfill should be adequately compacted. Machine compaction should not be used within 6 inches of the cable.

B. Plowing

1. Plowing in of cable in soil containing rock or other solid material should be done in such a manner that the solid material will not damage the cable, either during the plowing operation or afterward.

2. The design of cable plowing equipment and the plowing-in operation should be such that the cable will not be damaged by bending, side-wall pressure, or excessive cable tension.

C. Boring

Where a cable system is to be installed by boring and the soil and surface loading conditions are such that solid material in the region may damage the cable, the cable shall be adequately protected.

D. Depth of Burial

1. The distance between the top of a cable and the surface under which it is installed (depth of burial) shall be sufficient to protect the cable from injury or damage imposed by expected surface usage.

2. Burial depths as indicated in 353D2a are considered adequate, except as noted in b, c, or d following.

   a. Supply cables or conductors

Voltage	Depth of burial (in)
600 and below	24
601 to 22,000	30
22,001 to 40,000	36
40,001 and above	42

b. In areas where frost conditions could damage cables, greater burial depths than indicated above may be desirable.

c. Lesser depths than indicated above may be used where supplemental protection is provided.

d. Where the surface is not to final grade, under which a cable is to be installed, the cable should be placed so as to meet or exceed the requirements indicated above, both at the time of installation and subsequent thereto.

354. Random Separation — Additional Requirements

These rules apply to cables or conductors when the radial separation between them will be less than 12 in.

A. Supply Cables or Conductors
The cables or conductors of a supply circuit and those of another supply circuit may be buried together at the same depth with no deliberate separation between facilities, provided all parties involved are in agreement.

B. Communication Cables or Conductors
The cables or conductors of a communication circuit and those of another communication circuit may be buried together and at the same depth with no deliberate separation between facilities, provided all parties involved are in agreement.

C. Supply and Communication Cables or Conductors
Supply cables or conductors and communication cables or conductors may be buried together at the same depth with no deliberate separation between facilities, provided all parties involved are in agreement and the following requirements are met:

1. Voltage
   a. Grounded supply systems shall not be operated in excess of 22000 V to ground.
   b. Ungrounded supply systems shall not be operated in excess of 5300 V phase to phase.

2. Bare or Semiconducting Jacketed Grounded Conductor
   a. A supply facility operating above 300 V to ground shall include a bare or semiconducting jacketed grounded conductor in continuous contact with the earth. This conductor, adequate for the expected magnitude and duration of the fault current which may be imposed, shall be one of the following:

(1) a sheath, a shield, or both

(2) multiple concentric conductors closely spaced circumferentially

(3) a separate conductor in contact with the earth and in close proximity to the cable, where such cable or cables also have a grounded sheath or shield not necessarily in contact with the earth. The sheath, shield, or both, as well as the separate conductor, shall be adequate for the expected magnitude and duration of the fault currents which may be imposed.

NOTE: This is applicable when a cable in nonmetallic duct is considered as a direct buried cable installation and random separation is desired.

EXCEPTION: Where buried cable passes through a short section of conduit such as under a roadway, the contact with earth of the grounded conductor can be omitted, provided the grounded conductor is continuous through the conduit.

b. The bare conductor or conductors in contact with the earth shall be of suitable corrosion resistant material. The conductor covered by a semiconducting jacket shall be compatible with the jacketing compound.

c. The radial resistivity of the semiconducting jacket shall not be more than 20 meter ohms and shall remain essentially stable in service. The radial resistivity of the jacket material is that value calculated from measurements on a unit length of cable, of the resistance between the concentric neutral and a surrounding conducting medium. Radial resistivity is equal to the resistance of a unit length times the surface area of jacket divided by the average thickness of the jacket over the neutral conductors. All dimensions are to be expressed in meters.

3. Ungrounded Supply Systems
Cables of an ungrounded supply system operating above 300 V shall be of effectively grounded concentric shield construction in continuous contact with the earth. Such cables shall be maintained in close proximity to each other.

D. Multiple Cable Systems

More than one cable system buried in random separation may be treated as one system when considering clearance from other underground structures or facilities.

E. Protection

1. Supply circuits operating above 300 V to ground or 600 V between conductors shall be so constructed, operated, and maintained that when faulted, they shall be promptly de-energized initially or following subsequent protective device operation (phase-to-ground faults for grounded circuits, phase-to-phase faults for ungrounded circuits).

2. Ungrounded supply circuits operating above 300 V shall be equipped with a ground fault indication system.

3. Communication protective devices shall be adequate for the voltage and currents expected to be impressed on them in the event of contact with the supply conductors.

4. Adequate bonding shall be provided between the effectively grounded supply conductor or conductors and the communication cable shield or sheath at intervals which should not exceed 1000 ft.

5. In the vicinity of supply stations where large ground currents may flow, the effect of these currents on communication circuits should be evaluated before communication cables are placed in random separation with supply cables.

Section 36. Risers

360. General

A. Mechanical protection for supply conductors or cables shall be provided as required by Part 2 of this code. This protection should extend at least 1 ft below ground level.

B. Supply conductors or cable should rise vertically from the cable trench with only such deviation as necessary to permit a reasonable cable bending radius.

C. Exposed conductive pipes or guards containing supply conductors or cables shall be grounded in accordance with Rule 314.

361. Installation

A. The installation should be designed so that water does not stand in riser pipes above the frost line.

B. Conductors or cables shall be supported in a manner designed to prevent damage to conductors, cables, or terminals.

C. Where conductors or cables enter the riser pipe or elbow, they shall be installed in such a manner that shall minimize the possibility of damage due to relative movement of the cable and pipe.

362. Pole Risers — Additional Requirements

A. Risers should be located on the pole in the safest available position with respect to climbing space and possible exposure to traffic damage.

B. The number, size, and location of riser ducts or guards shall be limited to allow adequate access for climbing.

363. Pad-Mounted Installations

A. Supply conductors or cables rising from the trench to transformers, switchgear, or other equipment mounted on pads shall be so placed and arranged that they will not bear on the edges of holes through the pad nor the edges of bends or other duct work below the pad.

B. Cable entering pad-mounted equipment shall be maintained substantially at adequate depth for the voltage class until it becomes protected by being directly under the pad, unless other suitable mechanical protection is provided.

Section 37. Supply Cable Terminations

370. General

A. Cable terminations shall be designed and constructed to meet the requirements of Rule 333.

B. Riser terminations not located within a vault, pad-mounted equipment, or similar enclosure shall be installed in a manner designed to assure that clearance specified in Parts 1 and 2 of this code are maintained.

C. A cable termination shall be designed to prevent moisture penetration into the cable where such penetration is detrimental to the cable.

D. Where clearances between parts at different potentials are reduced below those adequate for the voltage and BIL (basic impulse insulation level), suitable insulating barriers or fully insulated terminals shall be provided to meet the required equivalent clearances.

371. Support at Terminations

A. Cable terminations shall be installed in a manner designed to maintain their installed position.

B. Where necessary, cable shall be supported or secured in a manner designed to prevent the transfer of damaging mechanical stresses to the termination, equipment, or structure.

372. Identification

Suitable circuit identification shall be provided for all terminations.

EXCEPTION: This requirement does not apply where the position of the termination, in conjunction with diagrams or maps supplied to workmen, gives sufficient identification.

373. Separations and Clearances in Enclosures or Vaults

A. Adequate electrical clearances and separations of supply terminations shall be maintained, both between conductors and between conductors and ground, consistent with the type of terminator used.

B. Where exposed live parts are in an enclosure, clearances and separations or insulating barriers adequate for the voltages and the design BIL shall be provided.

C. Where a termination is in a vault, uninsulated live parts are permissible provided they are guarded or isolated.

374. Grounding

A. All exposed conducting surfaces of the termination device, other than live parts and equipment to which it is attached, shall be effectively grounded, bonded, or both.

B. Conductive structures supporting cable terminations shall be effectively grounded.

EXCEPTION: Grounding, bonding, or both is not required where the above parts are isolated or guarded.

Section 38. Equipment

380. **General**

A. Equipment includes:
   1. Buses, transformers, switches, etc, installed for the operation of the electric-supply system.
   2. Repeaters, loading coils, etc, installed for the operation of the communication system.
   3. Auxiliary equipment such as sump pumps, convenience outlets, etc, installed incidental to the presence of the supply or communication systems.

B. Where equipment is to be installed in a joint-use manhole, it shall be done with the concurrence of all parties concerned.

C. Supporting structures, including racks, hangers, or pads and their foundations shall be designed to sustain all loads and stresses expected to be imposed by the supported equipment including those stresses caused by its operation.

381. **Design**

A. The expected thermal, chemical, mechanical, and environmental conditions at the location shall be considered in the design of all equipment and mountings.

B. All equipment, including auxiliary devices, shall be designed to withstand the effects of normal, emergency, and fault conditions expected during operation.

C. Switches shall be provided with clear indication of contact position, and the handles or activating devices clearly marked to indicate operating directions.

   RECOMMENDATION: The handles or control mechanism of all switches throughout the system should operate in a like direction to open and in a uniformly different direction to close in order to minimize errors.

D. Remotely controlled or automatic devices shall have provisions for local blocking to prevent operation if such operation may result in a hazard to the workman.

E. Enclosures containing fuses and interrupter contacts shall be designed to withstand the effects of normal, emergency, and fault conditions expected during operation.

F. When tools are to be used to connect or disconnect energized devices, space or barriers shall be designed to provide adequate clearance from ground or between phases.

G. Where pad-mounted equipment is not within a fenced or other protected area, access to contained exposed live parts in excess of 600 V shall require two separate procedures. The first procedure shall be the opening of a door or barrier which is locked or otherwise secured against unauthorized entry. The second access procedure shall be contingent on the completion of the first.

382. Location in Underground Structures

A. Equipment shall not obstruct personnel access openings in manholes or vaults nor shall it prevent easy egress by men working in the structures containing the equipment.

B. Equipment shall not be installed closer than 8 in to the back of fixed ladders and shall not interfere with the proper use of such ladders.

C. Equipment should be arranged in a manhole or vault to permit installation, operation, and maintenance of all items in such structures.

D. Switching devices which have provision for manual or electrical operation shall be operable from a safe position. This may be accomplished by use of portable auxiliary devices, temporarily attached.

E. Equipment should not interfere with drainage of the structure.

F. Equipment shall not interfere with the ability to ventilate any structure or enclosure.

383. Installation

A. Provisions for lifting, rolling to final position, and mounting shall be adequate for the weight of the device.

B. Live parts shall be guarded or isolated to prevent contact by persons in a normal position adjacent to the equipment.

C. Operating levers, inspection facilities, and test facilities shall be visible and readily accessible when equipment is in final location without moving permanent connections.

D. Live parts shall be isolated or protected from exposure to conducting liquids or other material expected to be present in the structure containing the equipment.

E.   Operating controls of supply equipment, readily accessible to unauthorized personnel, shall be secured by bolts, locks, or seals.

384. Grounding

A.   Cases and enclosures made of conductive material shall be effectively grounded or guarded.

B.   Guards constructed of conductive material shall be effectively grounded.

385. Identification

Where transformers, regulators, or other similar equipment operate in multiple, tags, diagrams, or other suitable means shall be used to indicate that fact.

Section 39. Installation in Tunnels

390. General

A.   The installation of supply and communication facilities in tunnels shall meet the applicable requirements contained elsewhere in Part 3 of this code as supplemented or modified by this section.

B.   Where the space occupied by supply or communications facilities in a tunnel is accessible to other than qualified persons, or where supply conductors do not meet the requirements of Part 3 of this code for cable systems, the installation shall be in accordance with the applicable requirements of Part 2 of this code.

C.   All parties concerned must be in agreement with the design of the structure and designs proposed for installations within it.

391. Environment

A.   When the tunnel is accessible to the public or when workmen must enter the structure to install, operate, or maintain the facilities in it, the design shall provide a controlled safe environment including where necessary, barriers, detectors, alarms, ventilation, pumps, and adequate safety devices for all facilities. Controlled safe environment shall include:

     1.   Design to avoid poisonous or suffocation atmosphere

2. Design to protect persons from pressurized lines, fire, explosion, and high temperatures

3. Design to avoid unsafe conditions due to induced voltages

4. Design to prevent hazards due to flooding

5. Design to assure egress; two directions for egress shall be provided for all points in tunnels

6. Working space, in accordance with Rule 323 B, the boundary of which shall be a minimum of 2 ft away from vehicular operating space or from exposed moving parts of machinery

7. Safeguards designed to protect workmen from hazards due to the operation of vehicles or other machinery in tunnels

8. Unobstructed walkways for workmen in tunnels.

B. A condition of occupancy in multiple-use tunnels by supply and communications facilities shall be that the design and installation of all facilities is coordinated to provide a safe environment for the operation of supply facilities, communications facilities, or both. Safe environment for facilities shall include:

1. Means to protect equipment from harmful effects of humidity or temperature

2. Means to protect equipment from harmful effects of liquids or gases

3. Coordinated design and operation of corrosion control systems

American National Standard

National Electrical Safety Code

Part 4 (Sections 40-43). Rules for the Operation of Electric-Supply and Communications Lines and Equipment

Secretariat
Institute of Electrical and Electronics Engineers

4

Section A. Reference to Other Standards

The following American National Standards, or the latest revisions thereof, approved by the American National Standards Institute should be used in conjunction with this standard where applicable.

ANSI A14.1-1975 and A14.1a-1977, Safety Requirements for Portable Wood Ladders

ANSI A14.2-1972 and A14.2a-1977, Safety Requirements for Portable Metal Ladders

ANSI A14.3-1974, Safety Requirements for Fixed Ladders

ANSI B15.1-1972, Safety Standard for Mechanical Power-Transmission Apparatus

ANSI A92.2-1979, Vehicle Mounted Elevating and Rotating Work Platforms

ANSI Z53.1-1978, Safety Color Code for Marking Physical Hazards

ANSI Z87.1-1979, Practice for Occupational and Educational Eye and Face Protection

ANSI Z88.2-1969, Practices for Respiratory Protection

ANSI Z89.2-1971, Safety Requirements for Industrial Protective Helmets for Electrical Workers — Class B

ANSI/ASTM D120-77, Standard Specification for Rubber Insulating Gloves

ANSI/ASTM D178-77, Standard Specification for Rubber Insulating Matting

ANSI/ASTM D1048-77, Standard Specification for Rubber Insulating Blankets

ANSI/ASTM D1049-79, Standard Specification for Rubber Insulating Covers

ANSI/ASTM D1050-77, Standard Specification for Rubber Insulating Line Hose

ANSI/ASTM F478-76, Standard Specification for In-Service Care of Insulating Line Hose and Covers

ANSI/ASTM F479-79, Standard Specification for In-Service Care of Insulating Blankets

ANSI/NFPA 10-1978, Portable Fire Extinguishers

NOTE

After ANSI C2.4-1973 was originally approved 30 June 1972, OSHA (Occupational Safety and Health Administration) issued 29CFR1926 Subpart V applying to employee safety in construction. The 1981 Edition revisions did not address the differences. Footnotes have been provided in the text where OSHA requirements conflict with ANSI C2-1981 requirements.

PART 4. Rules for the Operation of Electric-Supply and Communications Lines and Equipment

Section 40. Purpose and Scope

400. Purpose

The purpose of Part 4 of this code is to provide practical work rules as one of the means of safeguarding employees and the public from injury.

401. Scope

Part 4 of this code covers work rules to be followed in the installation, operation, and maintenance of electric supply and communication systems.

Section 41. Supply and Communications Systems — Rules for Employers

410. General Requirements

A. General

1. The employer shall inform each employee working on or about communications equipment or electric-supply equipment and the associated lines, of the safety rules governing the employee's conduct while so engaged. When deemed necessary, the employer shall provide a copy of such rules.

2. Employers shall utilize positive procedures to secure compliance with these rules. Cases may arise, however, where the strict enforcement of some particular rule could seriously impede the safe progress of the work at hand; in such cases the employee in charge of the work to be done may make such temporary modification of the rules as will accomplish the work without increasing the hazard.

3. If a difference of opinion arises with respect to the application of these rules, the decision of the employer or his authorized agent shall be final. This decision shall not result in any employee performing work in a manner which is unduly hazardous to himself or to his fellow workers.

B. Emergency Procedures and First Aid Rules

1. Employees shall be informed of procedures to be followed in case of emergencies and rules for first aid, including approved methods of resuscitation. Copies of such procedures and rules should be kept in conspicuous locations in vehicles and places where the number of employees and the nature of the work warrants.

2. Employees working on communications or electric-supply equipment or lines shall be regularly instructed in methods of first aid and emergency procedures, if their duties warrant such training.

C. Responsibility

1. A qualified system operator or other designated authority shall be in charge of the operation of the equipment and lines and shall be responsible for their safe operation.

2. If more than one person is engaged in work on or about the same equipment or line, one person shall be designated as in charge of the work to be performed. Where there are separate work locations one person may be designated at each location.

411. Protective Methods and Devices

A. Methods

1. Access to rotating or energized equipment shall be restricted to authorized personnel.

2. Diagrams, showing plainly the arrangement and location of the electric-supply equipment and lines, shall be maintained on file and readily available to authorized personnel for that portion of the system for which they are responsible.

3. Employees shall be instructed as to the character of the equipment or lines and methods to be used before any work is undertaken thereon.

4. Employees should be instructed to take additional precautions to insure their safety when conditions create unusual hazards.

B.　Devices and Equipment

An adequate supply of protective devices and equipment, sufficient to enable employees to meet the requirements of the work to be undertaken, and first aid equipment and materials shall be available in readily accessible and, where practical, conspicuous places.

NOTE: The following is a list of some common protective devices and equipment, the number and kinds of which will depend upon the requirements of each case:

1. Insulating wearing apparel such as rubber gloves, rubber sleeves, and headgear
2. Insulating shields, covers, mats, and platforms
3. Insulating tools for handling or testing a energized equipment or lines
4. Protective goggles
5. *Men at work* tags, portable danger signs, traffic cones, and flashers
6. Body belts and safety straps
7. Fire extinguishing equipment designed for safe use on energized parts or plainly marked that they must not be so used
8. Protective grounding materials and devices
9. Portable lighting equipment
10. First aid equipment and materials

C.　Inspection and Testing of Protective Devices

1. Protective devices and equipment shall be periodically inspected or tested to insure that they are in safe working condition.
2. Insulating gloves, sleeves, and blankets shall be inspected before use. Insulating gloves and sleeves shall be tested periodically and as frequently as their use requires.
3. Body belts, safety straps, and other personal equipment, whether furnished by employer or employee, shall be inspected from time to time to insure that they are in safe working condition.

D.　Warning Signs

Permanent warning signs shall be displayed in conspicuous places at all entrances to electrical-supply stations, substations, and other enclosed walk-in areas containing exposed current-carrying parts.

E.　Identification of Supply Circuits

Means shall be provided so that identification of supply circuits can be determined before work is undertaken.

Section 42.   Supply Systems — Rules for Employees

420. General Precautions

A.   Rules and Emergency Methods

The safety rules shall be carefully read and studied. Employees may be called upon at any time to show their knowledge of the rules.

Employees shall familiarize themselves with approved methods of first aid, rescue techniques, and fire extinguishment.

B.   Safeguarding Oneself and Others

The care exercised by others should not be relied upon for protection.

1.   Employees shall heed warning signs and signals and warn others who are in danger near energized equipment or lines.

2.   Employees shall report promptly to the proper authority any of the following:

   a.   Line or equipment defects such as abnormally sagging wires, broken insulators, broken poles, or lamp supports.

   b.   Accidentally energized objects such as conduits, light fixtures, or guys.

   c.   Other defects that may cause a dangerous condition.

3.   Employees whose duties do not require them to approach or handle electric equipment and lines shall keep away from such equipment or lines and should avoid working in areas where objects and materials may be dropped by persons working overhead.

4.   Employees who work on or near energized lines shall consider all of the effects of their actions, taking into account their own safety as well as the safety of other employees on the job site, or on some other part of the affected electric system, the property of others, and the public in general.

C.   Qualifications of Employees

1.   Inexperienced employees working on or about energized equipment or lines shall work under the direction of an experienced and qualified person at the site.

2. Employees who do not normally work on or about electric-supply lines and equipment but whose work brings them into these areas for certain tasks shall proceed with this work only when authorized by a qualified person.

3. If an employee is in doubt as to the safe performance of any work assigned to him, he shall request instructions from his supervisor or other qualified person.

D. **Energized or Unknown Conditions**

Electric-supply equipment and lines shall be considered energized, unless they are positively known to be de-energized. Before starting work, preliminary inspections or tests shall be made to determine existing conditions. Operating voltages of equipment and lines should be known before working on or near energized parts.

E. **Ungrounded Metal Parts**

All ungrounded metal parts of equipment or devices such as transformer cases and circuit breaker housings shall be considered to be energized at the highest voltage to which they are exposed, unless these parts are known by test to be free from such voltage.

F. **Arcing Conditions**

Employees should keep all parts of their bodies as far away as practical from switches, circuit breakers, or other parts at which arcing may occur during operation.

G. **Batteries**

1. Enclosed areas containing storage batteries shall be adequately ventilated. Smoking, the use of open flames, and tools which may produce sparks should be avoided in such enclosed areas.

2. Employees shall use eye and skin protection when handling an electrolyte.

3. Employees shall not handle energized parts of batteries unless necessary precautions are taken to avoid shock and short circuits.

H. **Tools and Protective Equipment**

Employees shall use the personal protective equipment, the protective devices, and the special tools provided for their work. Before starting work, these devices and tools shall be carefully inspected to make sure that they are in good condition.

I. Clothing

 1. The clothing worn by an employee in the performance of his duties shall be suitable for the work to be performed and the conditions under which such work is to be performed.

 2. When working in the vicinity of energized lines or equipment, the wearing of exposed metal articles such as key or watch chains, rings, wrist watches, bands, or zippers should be avoided.

J. Supports and Ladders

 1. No employee, or any material or equipment, shall be supported or permitted to be supported on any portion of a tree, pole structure, scaffold, ladder, walkway, or other elevated structure or aerial device, etc, without it first being determined, to the extent practical, that such support is adequately strong, in good condition, and properly secured in place.

 2. Portable wood ladders intended for general use shall not be painted except with a clear nonconductive coating, nor shall they be longitudinally reinforced with metal.

 3. Portable metal ladders intended for general use shall not be used when working on or near energized parts.

 4. If portable ladders are made partially or entirely conductive for specialized work, necessary precautions shall be taken to insure that their use will be restricted to the work for which they are intended.

K. Safety Straps

 1. An employee working in an elevated position shall use a suitable safety strap or other approved means to prevent falling.

 2. Safety traps or other devices shall be inspected by the employee to assure that they are in safe working condition.

 3. Before an employee trusts his weight to the safety strap or other device, he shall determine that the snaps or fastenings are properly engaged and that he is secure in his body belt and safety strap.

L. Fire Extinguishers

In fighting fires near exposed energized parts, employees shall use fire extinguishers or materials which are suitable for the purpose. If this is not possible, all adjacent and affected equipment should first be de-energized.

M. **Repeating Messages**

Each employee receiving an oral message concerning the switching of lines and equipment shall immediately repeat it back to the sender and obtain his identity. Each employee sending such an oral message shall require it to be repeated back to him by the receiver and secure the latter's identity.

N. **Machines and Moving Parts**

Employees working on normally moving parts of remotely controlled equipment shall be protected against accidental starting by proper tags installed on the starting devices, and by locking or blocking where practical. Employees shall, before starting any work, satisfy themselves that these protective devices have been installed. When working near automatically or remotely operated equipment such as circuit breakers which may operate suddenly, employees shall avoid being in a position where they might be injured from such operation.

O. **Fuses**

When fuses must be installed or removed with one or both terminals energized above 1 kV, special tools insulated for the voltage shall be used. Insulating tools or gloves should be used for voltages between 300 and 1000. When installing expulsion type fuses, employees shall wear safety glasses or safety goggles and take precautions to stand clear of the exhaust path of the fuse barrel.

P. **Cable Reels**

Cable reels shall be securely blocked so they cannot roll accidentally.

421. **Operating Routines**

A. **Duties of a System Operator**

A system operator shall:

1. Keep informed of operating conditions affecting the safe and reliable operation of the system.
2. Maintain a suitable record showing operating changes in such conditions.

B. **Duties of a Foreman**

A foreman shall:

1. Adopt such precautions as are within his power to prevent accidents and to see that the safety rules and operating procedures are observed by the employees under his direction.

2. Make all the necessary records and report to the system operator when required.

3. As far as possible, prevent unauthorized persons from approaching places where work is being done.

4. Prohibit the use of any tools or devices unsuited to the work at hand or which have not been tested or inspected as required by these rules.

C. Guides

Persons accompanying uninstructed employees or visitors near electric equipment or lines shall be qualified to safeguard the people in their care and see that the safety rules are observed.

D. Authorization

1. Specific Work

Authorization from the system operator shall be secured before work is begun on or about station equipment, transmission, or interconnected feeder circuits and where circuits are to be de-energized at stations. The system operator shall be notified when such work ceases.

EXCEPTION: In an emergency, to protect life or property, or when communication with the system operator is difficult, because of storms or other causes, any qualified employee may make repairs on or about the equipment or lines covered by this rule without special authorization if he can clear the trouble promptly with available help in compliance with the remaining rules. The system operator shall thereafter be notified as soon as possible of the action taken.

2. Operations at Stations

In the absence of specific operating schedules, employees shall secure authorization from the system operator before opening and closing supply circuits or starting and stopping equipment affecting system operation at stations.

EXCEPTION: In an emergency, to protect life or property, any qualified employee may open circuits and stop moving equipment without special authorization if, in his judgement, his action will promote safety, but the system operator shall be notified as soon as possible of such action, with reasons therefor.

3. De-energizing Sections of Circuits by Sectionalizing Devices

Employees shall obtain authorization from the system operator before de-energizing sections of circuits.

EXCEPTION: Sections of distribution circuits are excepted if the system operator is notified as soon as possible after the action is taken.

E. Re-energizing After Work

Instructions to re-energize equipment or lines which have been de-energized by permission of the system operator shall not be issued by him until all employees who requested the line to be de-energized have reported clear. Employees who have requested equipment or lines de-energized for other employees or crews shall not request that equipment or lines to be re-energized until all of the other employees or crews have reported clear. The same procedure shall be followed when more than one location is involved.

F. Tagging Electric-Supply Circuits

Equipment or circuits that are to be treated as de-energized shall have suitable tags attached to all points where such equipment or circuits can be energized. Controls that are to be de-activated during the course of work on energized or de-energized equipment or circuits also shall be tagged. The tags shall be placed to identify plainly the equipment or circuits on which work is being performed.

G. Maintaining Service

1. Closing Tagged Circuits Which Have Opened Automatically

    When controls upon which tags have been placed open automatically, they shall be left open until re-closing has been authorized.

2. Closing Circuits Opened Automatically

    When circuits open automatically, local operating rules shall determine in what manner and how many times they may be closed with safety.

3. Unintentional Grounds on Delta Circuits

    Unintentional grounds on delta circuits shall be removed as soon as practical.

H. Area Protection

1. Vehicular and Pedestrian Traffic

    a. Before engaging in work that may endanger the public, warning signs or traffic control devices, or both, shall be placed conspicuously to approaching traffic. Where further protection is needed, suitable barrier guards shall be erected. Where the nature of work and traffic requires it, a man shall be stationed to warn traffic while the hazard exists.

    b. In case openings or obstructions in the street, sidewalk, walkways, or on private property are being worked on or left unattended during the day, danger signals, such as warning signs and flags, shall be effectively displayed. Under these same conditions at night, warning lights shall be prominently displayed and excavations shall be enclosed with protective barricades.

2. Employees
    a. If the work exposes energized or moving parts that are normally protected, danger signs shall be displayed and suitable guards erected to warn other personnel in the area.
    b. When working in one section where there is a multiplicity of such sections, such as one panel of a switchboard, one compartment of several, or one portion of a substation, personnel shall mark the work area conspicuously and place barriers to prevent accidental contact with energized parts in that section or adjacent sections.

3. Crossed or Fallen Wires
An employee finding any crossed or fallen wires which are or may create a hazard shall remain on guard or initiate appropriate action. If the employee can observe the rules for handling energized parts by the use of insulating equipment, he may correct the condition at once; otherwise he shall first secure authorization for so doing. (See Rule 421 D for special authorization.)

I. **Protecting Employees by Switches and Disconnectors**
When equipment or lines are to be disconnected from any source of electric energy for the protection of employees, the operator shall first open the switches or circuit breakers designated for operation under the load involved.

422. Handling Energized Equipment or Lines

A. **General Requirements**
1. Work on Energized Lines and Equipment
When working on energized lines and equipment, one of the following safeguards shall be applied:
    a. Insulate employee from energized parts.
    b. Isolate or insulate employee from ground and grounded structures, and potentials other than the one being worked on.

2. Covered (Noninsulated) Wire

Employees should not place dependence for their safety on the covering of wires. All precautions in this section for handling energized parts shall be observed in handling covered wires.

3. Exposure to Higher Voltages

Every employee working on or about equipment or lines exposed to voltages higher than those guarded against by the safety appliances provided shall assure himself that the equipment or lines on which he is working are free from dangerous leakage or induction or have been effectively grounded.

4. Cutting Into Insulating Coverings of Energized Conductors

When the insulating covering on energized wires or cable must be cut into, the employee shall use suitable tools. While doing such work, suitable eye protection and insulating gloves with protectors shall be worn. Employees shall exercise extreme care to prevent short-circuiting conductors when cutting into the insulation.

5. Metal Tapes or Ropes

Metal measuring tapes, and tapes or ropes containing metal threads or strands, shall not be used closer to exposed energized parts than the distance specified in Rule 422 B. Also, care should be taken when extending metallic ropes or tapes parallel to and in the proximity of high voltage lines because of the effect of induced voltages.

6. Work Equipment or Material Extending Into Energized Areas

Equipment or material of a noninsulating substance which is not bonded to an effective ground and extends into an energized area, and could approach energized equipment closer than the distance specified in Rule 422 B, shall be treated as though it is energized at the same voltage as the line or equipment to which it is exposed.

B. Clearance from Live Parts

No employee shall approach or take any conductive object without a suitable insulating handle within the distances of any exposed energized part listed in Table 1A unless he is insulated from the energized part, the energized part is insulated from the employee, or the employee is insulated from all conducting surfaces other than the one upon which he is working. (See Tables 1 and 2.) Gloves rated for the voltage involved shall be considered effective insulation of the employee from the energized part. (For bare-hand live-line work, see Rule 427.)

C. Requirement for Assisting Employee
In inclement weather or at night no employee shall work alone outdoors on or dangerously near energized conductors or parts of more than 750 V between conductors.

EXCEPTION: This shall not preclude a qualified employee, working alone, from cutting trouble in the clear, switching, replacing fuses, or similar work if such work can be performed safely.

D. When to De-energize Parts
An employee shall not approach, or willingly permit others to approach, any exposed ungrounded part normally energized except as permitted by Rule 422 B, unless the supply equipment or lines are de-energized.

Table 1A
AC Minimum Clearance from Live Parts

Nominal voltage in kilovolts phase to phase	Distance in feet phase to employee
1 to 34.5†	2
46	2½
69	3
115†	3
138	3½
161†	3½
230	5

NOTE: These distances take into consideration the highest switching surge an employee will be exposed to on any system with live-line tools as the insulating material and maximum operating voltages as stated in ANSI C84.1-1970.

†OSHA 1926.950 specifies minimum clearances greater than ANSI C2.4 as follows: 15.1—35 kV: 2 ft, 4 in; 115 kV: 3 ft, 4 in; 161 kV: 3 ft, 8 in.

Table 1B
AC Minimum Clearance with Surge Factor*‡

Maximum switching surge factor	Distance in feet					
	Air			Live-line tool		
	Nominal/maximum phase-to-phase kilovolts					
	345/ 362	500/ 550	700/ 765	345/ 362	500/ 550	700/ 765
1.5 or below	5/5	5/5	7/8½	5/5	5/5½	8/9
1.6	5/5	5/5½	8/9	5/5	5½/6	8½/10
1.7	5/5	5/6	8½/10	5/5	5½/6½	9½/10½
1.8	5/5	5½/6½	9½/11	5/5	6½/7	10½/12
1.9	5/5	6/7	10½/12	5/5	7/8	11/13
2.0	5/5	6½/7½	11½/13	5/5	7½/8½	12/14
2.1	5/5	7/8½	12/14	5/5	8/9	13/15
2.2	5/5	7½/9	13/15½	5/5½	8½/9½	14/16½
2.3	5/5	8½/9½		5/5½	9/10½	
2.4	5/5½	9/10		5½/6	9½/11	
2.5	5½/6	9½/11		6/6½	10/12	
2.6	5½/6			6/6½		
2.7	6/6½			6½/7		
2.8	6½/7			7/7½		

*The distances specified in this table may be applied *only* where the operating voltage and the switching surge factor are known and are supplied by the employer.

NOTE: On systems where the voltage is controlled to operate at a maximum value in between the nominal/maximum listed in the table, clearance distances may be determined by interpolation, rounding to the next higher ½ ft.

---

‡There is no provision in OSHA 1926 for clearance reduction through application of switching surge factors.

Table 2A
DC Minimum Clearance from Live Parts

Maximum voltage conductor to ground kilovolts	Distance in feet
250	3½
400	6
500	8½
750	16

NOTE: These distances are based on the highest transient overvoltage an employee will be exposed to on any system with live-line tools as the insulating material.

Table 2B
DC Minimum Clearance with Overvoltage Factor

Transient overvoltage factor	Distance in feet							
	Air				Live-line tool			
	Maximum conductor to ground kilovolts							
	250	400	500	750	250	400	500	750
1.5 or below	3	4½	6	11	3	5	6½	12
1.6	3	4½	6½	12	3½	5	7	13
1.7	3	5	7	13½	3½	5½	8	14½
1.8	3½	5½	7½	15	3½	6	8½	16

*The distances specified in this table may be applied *only* where the transient overvoltage factor is known and is supplied by the employer.

E. Opening and Closing Switches

Manual switches and disconnectors shall always be closed by a continuous motion. Care should be exercised in opening switches to avoid serious arcing.

F. Working Position

Employees should avoid working on equipment or lines in any position from which a shock or slip will tend to bring the body toward exposed parts at a potential different than the employee's body. Work should, therefore, generally be done from below, rather than from above.

G. Making Connections

In connecting de-energized equipment or lines to an energized circuit by means of a conducting wire or device, employees should first attach the wire to the de-energized part. When disconnecting, the source end should be removed first. Loose conductors should be kept away from exposed energized parts.

H. Current Transformer Secondaries

The secondary of a current transformer shall not be opened while energized. If the entire circuit cannot be properly de-energized before working on an instrument, a relay, or other section of a current transformer secondary circuit, the employee shall bridge the circuit with jumpers so that the current transformer secondary will not be opened.

I. Capacitors

Before employees work on capacitors, the capacitors shall be disconnected from the energizing source, short-circuited, and grounded. Any line to which capacitors are connected shall be short-circuited and grounded before it is considered de-energized. Since capacitor units may be connected in series-parallel, each unit shall be shorted between all insulated terminals and the capacitor tank before handling. Where the tanks of capacitors are on ungrounded racks, the racks shall also be grounded. The internal resistor shall not be depended upon to discharge capacitors.

423. De-Energizing Equipment or Lines to Protect Employees

A. Application of Rule

1. When employees must depend on others for operating switches to de-energize circuits on which they are to work, or must secure special authorization before them-

selves operating such switches, the precautionary
measures that follow shall be taken in the order given
before work is begun.

2. If the employee under whose direction a section of a
circuit is disconnected is in sole charge of the section and
of the means of disconnection, those portions of the
measures that follow which pertain to his dealing with
the system operator may be omitted.

B. Employee's Request

The employee in charge of the work must apply to the
system operator to have the particular section of equipment
or lines de-energized, identifying it by position, letter, color,
number, or other means.

C. Opening Disconnectors and Tagging

The system operator shall direct the opening of all switches
and disconnectors through which electric energy may be
supplied to the particular section of equipment and lines to
be de-energized, and shall direct that such switches and
disconnectors be rendered inoperable, where practical, and
tagged plainly indicating that men are at work. All automatically and remotely controlled switches shall also be tagged
at the point of control and should be rendered inoperable
where practical. A record shall be made when placing the tag
giving the time of disconnection, the name of the man
making the disconnection, the name of the employee who
requested the disconnection, and the name of the system
operator.

D. Employee's Protective Grounds

When all the switches and disconnectors designated have
been opened, rendered inoperable, where practical, and
tagged in accordance with Rule 423 C, and the employee has
been given permission to work by the system operator, the
employee in charge should immediately proceed to make his
own protective grounds or verify that adequate grounds have
been applied (see Rule 424) on the disconnected lines or
equipment. Such grounds shall be made between the point at
which work is to be done and every source of energy and as
close as practical to the work location. Where the making of
a ground is impractical or the conditions resulting therefrom
would be more hazardous than working on the lines without
grounding, the ground may be omitted by special permission
of the proper authority.

E. Proceeding with Work

After the equipment or lines have been de-energized and grounded, the employee in charge and those under his direction may proceed with work on the de-energized parts. Care must be taken to guard against adjacent energized circuits or parts.

F. Procedure for Other Crews

Each additional employee in charge desiring the same equipment or lines to be de-energized for the protection of himself, or the men under his direction, shall follow these procedures to secure similar protection.

G. Reporting Clear — Transferring Responsibility

The employee in charge, upon completion of his work, and after assuring himself that all men under his direction are in the clear, shall remove his protective grounds and shall report to the system operator that all tags protecting him may be removed.

The employee in charge who received the permission to work may transfer this permission and the responsibility for men under him as follows:

He shall personally inform the system operator of the proposed transfer, and if this is permitted, the men under his direction shall be notified and the name of the successor shall be entered at that time on the records. Thereafter the successor shall report clear and shall be responsible for the safety of all employees protected by the clearance.

H. Removal of Tags

The system operator shall then direct the removal of tags and the removal shall be reported back to him immediately by the persons removing them. Upon the removal of any tag, there shall be added to the record containing the name of the system operator and the person who requested the tag, the name of the person requesting removal, the time of removal, and the name of the person removing the tags. The name of the person requesting removal shall be the same as the name of the person requesting placement unless responsibility has been transferred according to Rule 423 G.

I. Restoring Service

Only after all protective grounds have been removed and after all protective tags have been removed by the above procedure from all points of disconnection may the system operator direct the closing of disconnectors and switches.

424. Protective Grounds

A. Installing Grounds

When placing temporary protective grounds on a normally energized circuit, the following precautionary measures shall be observed.

1. Size of Grounds

    The grounding device shall be of such size as to carry the maximum fault current that could flow at the point of grounding for the time necessary to clear the line.

2. Ground Connections

    The employee making a protective ground on equipment or lines shall first connect one end of the grounding device to an effective ground connection.

3. Test of Circuit

    The de-energized conductors and equipment which are to be grounded shall next be tested for voltage except where previously installed grounds are clearly in evidence. The employee shall keep all portions of his body at the required distance by using insulating handles of proper length or other suitable devices.

4. Completing Grounds

    If the test shows no voltage, or the local operating rules so direct, the free end of the grounding device shall next be brought into contact with the de-energized part using insulating handles or other suitable devices and securely clamped or otherwise secured thereto. Where bundled conductor lines are being grounded, adequate grounding of each subconductor should be made. Only then may the employee come within the distances from the de-energized parts specified in Rule 422 B or proceed to work upon the parts as upon a grounded part.

    NOTE: In stations, switches may be employed to connect the equipment or lines being grounded to the actual ground connection.

B. Removing Grounds

The employee shall first remove the grounding device from the de-energized parts using insulating handles or other suitable devices.

425. Overhead Lines

Employees working on or with overhead lines shall observe the following rules in addition to applicable rules contained elsewhere in Section 42.

A.   Checking Structures Before Climbing

1. Before climbing poles, ladders, scaffolds, or other elevated structures, employees shall determine, to the extent practical, that the structures are capable of sustaining the additional or unbalanced stresses to which they will be subjected.

2. Where there are indications that poles and structures may be unsafe for climbing, they shall not be climbed until made safe by guying, bracing, or other means.

B. Installing and Removing Wires or Cables

1. Precautions shall be taken to prevent wires or cables being installed or removed from contacting energized wires or equipment. Wires or cable which are not bonded to an effective ground and are being installed or removed in the vicinity of energized conductors shall be considered as being energized.

2. Sag of wires or cable being installed or removed shall be controlled to prevent danger to pedestrian and vehicular traffic.

3. Before installing or removing wire or cable, strains to which poles and structures will be subjected shall be considered and necessary action taken to prevent failure of supporting structures.

4. Employees should avoid contact with moving winch lines, especially near sheaves, blocks, and take-up drums.

C. Setting, Moving, or Removing Poles in or Near Energized Lines

1. When setting, moving, or removing poles in or near energized lines, precautions shall be taken to avoid direct contact of the pole with the energized conductors. Employees shall wear suitable insulating gloves or use other suitable means where voltages may exceed rating of gloves in handling poles where conductors energized at potentials above 750 V can be contacted. Employees performing such work shall not contact the pole with uninsulated parts of their body.

2. Contact with trucks, derricks, or other equipment which are not bonded to an effective ground and are being used to set, move, or remove poles in or near energized lines shall be avoided by employees standing on the ground or in contact with grounded objects unless the employee is wearing suitable protective equipment.

426. Underground Lines

Employees working on or with underground lines shall observe the following rules in addition to applicable rules contained elsewhere in Section 42.

A. **Guarding Manhole and Street Openings**

When covers of manholes, handholes, or vaults are removed, the opening shall be promptly protected with a barrier, temporary cover, or other suitable guard.

B. **Ventilation and Testing for Gas in Manholes and Unventilated Vaults**

1. The atmosphere shall be tested for flammable gas before entry.
2. Where flammable gases are detected, the work area shall be ventilated and made safe before entry.
3. Provision shall be made for an adequate continuous supply of air.

C. **Attendant on Surface**

While personnel are in a manhole, an employee shall be available on the surface in the immediate vicinity primarily to render assistance from the surface. This shall not preclude the employee on the surface from entering the manhole to provide short-term assistance.

EXCEPTION: This shall not preclude a qualified employee, working alone, from entering a manhole where energized cables or equipment are in service, for the purpose of inspection, housekeeping, taking readings, or similar work if such work can be performed safely.

D. **Flames**

1. Employees shall not smoke in manholes.
2. Where open flames must be used in manholes, extra precautions shall be taken to ensure adequate ventilation.
3. Before using open flames in an excavation in areas where combustible gases or liquids may be present, such as near gasoline service stations, the atmosphere of the excavation shall be tested and found safe or cleared of the combustible gases or liquids.

E. **Excavation**

1. Cables and other buried utilities in the immediate vicinity shall be located, to the extent practical, prior to excavating.
2. Hand tools used for excavating near energized supply

cables should be equipped with handles made of nonconductive material.

3. Mechanized equipment should not be used to excavate in
close proximity to cables and other buried utilities.

4. If a gas or fuel line should be broken or damaged,
employees shall:
   a. Leave the excavation open.
   b. Extinguish flames which could ignite the escaping gas
   or fuel.
   c. Notify the proper authority.
   d. Keep the public away until the condition is under
   control.

F. Identification

1. When underground facilities are exposed (electric, gas,
water, telephone, etc) they should be identified and shall
be protected as necessary to avoid damage.

2. Where multiple cables exist in an excavation, cables other
than the one being worked on shall be protected as
necessary.

3. Before cutting into a cable or opening a splice, the cable
should be identified and verified to be the proper cable.

4. A cable to be worked on as de-energized which cannot be
positively identified or determined to be de-energized
shall be pierced or servered with an approved tool at the
work location.

5. Before cutting into an energized cable the operating
voltage shall be determined and appropriate precautions
taken for handling conductors at that voltage.

G. Operation of Power Driven Equipment
Employees should avoid being in manholes where power
driven rodding equipment is in operation.

427. Bare-Hand Live-Line Work

All employees using bare-hand live-line work practices shall
observe the following rules in addition to applicable rules
contained elsewhere in Section 42.

A. Training
Employees shall be trained in bare-hand live-line work
methods before being permitted to use this technique on
energized lines.

B. Equipment
1. Insulated aerial devices used in bare-hand work shall be

tested before the work is started to assure the integrity of the insulation.

2. Insulated aerial devices and other equipment used in this work shall be maintained in a clean condition.
3. Tools and equipment shall not be used in a manner that will reduce the overall insulating strength of the insulated aerial device.

Table 3A
AC Minimum Clearance Live-Line Bare-Hand Work

Nominal voltage in kilovolts phase to phase	Distance in feet	
	Phase to ground	Phase to phase
1 to 34.5 §	2	2
46	2½	3
69 §	2½	3
115 §	3	4
138	3½	5
161 §	3½	5
230 §	5	8
345 §	7	13
500	11	21½
700	15½	31

NOTE: These distances take into consideration the highest switching surge an employee will be exposed to on any system with air as the insulating medium and maximum operating voltages as stated in ANSI C84.1-1970.

§ OSHA 1926.955 specifies minimum clearances greater than ANSI C2.4 as follows:

kV	To ground		To phase	
	ft	in	ft	in
15.1-35	2	4	2	4
69	3	0	—	—
115	3	4	—	—
161	3	8	5	6
230	—	—	8	4
345	—	—	13	4

Table 3B
AC Minimum Clearance with Surge Factor
Live-Line Bare-Hand Work*‡

Maximum switching surge factor	Distance in feet					
	Phase to ground			Phase to phase		
	Nominal/maximum phase-to-phase kilovolts					
	345/362	500/550	700/765	345/362	500/550	700/765
1.5 or below	5/5	5/5	7/8½	5/5	8/9½	13½/16
1.6	5/5	5/5½	8/9	5/5½	9/10½	15½/18
1.7	5/5	5/6	8½/10	5½/6	9½/11½	17/19½
1.8	5/5	5½/6½	9½/11	6/6½	10½/12½	19/22
1.9	5/5	6/7	10½/12	6½/7	11½/13½	20½/24
2.0	5/5	6½/7½	11½/13	7/7½	12½/15	22½/26
2.1	5/5	7/8½	12/14	7½/8	13½/16	24½/28½
2.2	5/5	7½/9	13/15½	8/9	15/17½	26½/31
2.3	5/5	8½/9½		9/9½	16/19	
2.4	5/5½	9/10		9½/10	17½/20	
2.5	5½/6	9½/11		10/10½	18½/21½	
2.6	5½/6			10½/11½		
2.7	6/6½			11½/12		
2.8	6½/7			12/13		

*The distances specified in this table may be applied *only* where the operating voltage and the switching surge factor are known and are supplied by the employer.

NOTE: On systems where the voltage is controlled to operate at a maximum value in between the nominal/maximum listed in the table, clearance distances may be determined by interpolation, rounding to the next higher ½ ft.

‡There is no provision in OSHA 1926 for clearance reduction through application of switching surge factors.

C. Clearances

The distances specified in Table 3A or 3B shall be maintained from all grounded objects and from lines and equipment at a different potential than that to which the insulated aerial device is bonded.

D. Bonding and Shielding

1. A conductive bucket liner or other suitable conducting device shall be provided for bonding the insulated aerial device to the energized line or equipment.

2. The employee shall be bonded to the insulated aerial device by use of conducting shoes, leg clips, or other suitable means.

3. Adequate electrostatic shielding shall be provided and used where necessary.

4. Before the employee contacts the energized part to be worked on, the aerial device shall be bonded to the energized conductor by means of a positive connection.

Section 43. Communications Systems — Rules for Employees

430. General Precautions

A. Rules and Emergency Methods

The safety rules shall be carefully read and studied. Employees may be called upon at any time to show their knowledge of the rules.

Employees shall familiarize themselves with approved methods of first aid, rescue techniques, and fire extinguishment.

B. Safeguarding Oneself and Others

The care exercised by others should not be relied upon for protection.

1. Employees shall heed warning signs and signals and warn others who are in danger near energized equipment or lines.

2. Employees shall report promptly to the proper authority any of the following:

    a. Line or equipment defects such as abnormally sagging wires, broken insulators, broken poles, or lamp supports.

    b. Accidentally energized objects such as conduit, light fixtures, or guys.

    c. Other defects that may cause a dangerous condition.

  3. Employees whose duties do not require them to approach or handle electric equipment and lines shall keep away from such equipment or lines and should avoid working in areas where objects and materials may be dropped by persons working overhead.

  4. Employees who work near energized supply lines shall consider all the effects of their actions, taking into account their own safety as well as the safety of other employees on the job site, the property of others, and the public in general.

C. Qualifications of Employees

  1. Inexperienced employees working in the vicinity of energized electric-supply equipment or lines shall work under the direction of an experienced and qualified person at the site.

  2. Employees who do not normally work in the vicinity of electric-supply lines and equipment but whose work brings them into these areas for certain tasks, shall proceed with this work only when authorized by a qualified person.

  3. If an employee is in doubt as to the safe performance of any work assigned to him, he shall request instructions from his supervisor or other qualified person.

D. Energized or Unknown Conditions

Electric-supply equipment and lines shall be considered to be energized unless they are known to be de-energized. Operating voltages of such equipment and lines should be known before working near energized parts.

E. Ungrounded Metal Parts

All ungrounded metal parts of equipment such as transformer cases and circuit breaker housings shall be considered energized at the highest voltage to which they are exposed, unless these parts are known by test to be free from such voltage.

F. Arcing Conditions

Employees should keep all parts of their bodies as far away as practical from brushes, commutators, switches, circuit breakers, or other parts at which arcing may occur during operation or handling.

G. Batteries

1. Enclosed areas containing storage batteries shall be adequately ventilated. Smoking, the use of open flames, and tools which may produce sparks should be avoided in such enclosed areas.

2. Employees shall use eye and skin protection when handling an electrolyte.

3. Employees shall not handle energized parts of batteries unless necessary precautions are taken to avoid shock and short circuits.

H. Tools and Protective Equipment

Employees shall use the personal protective equipment, the protective devices, and the special tools provided for their work. Before starting work, these devices and tools shall be carefully inspected to make sure that they are in good condition.

I. Clothing

The clothing worn by an employee in the performance of his duties shall be suitable for the work to be performed and the conditions under which such work is to be performed.

J. Supports and Ladders

1. No employee, or any material or equipment, shall be supported or permitted to be supported on any portion of a tree, pole structure, scaffold, ladder, walkway, or other elevated structure or aerial device, etc, without it first being determined, to the extend practical, that such support is adequately strong, in good condition, and properly secured in place.

2. Portable wood ladders intended for general use shall not be painted except with a clear nonconductive coating, nor shall they be longitudinally reinforced with metal.

3. Portable metal ladders shall not be used when working near energized parts of electric-supply systems.

4. If portable ladders are made partially or entirely conductive for specialized work, necessary precautions shall be taken to insure that their use will be restricted to the work for which they are intended.

K. Safety Straps

1. An employee working in an elevated position shall use a suitable safety strap or other approved means to prevent falling.

2. Safety straps or other devices shall be inspected by the employee to assure that they are in safe working condition.

3. Before an employee trusts his weight to the safety strap or other device, he shall determine that the snaps or fastenings are properly engaged and that he is secure in his body belt and safety strap.

L. Fire Extinguishers

In fighting fires near exposed energized parts of electric-supply systems, employees shall use fire extinguishers or materials which are suitable for the purpose. If this is not possible, all adjacent and affected equipment should first be de-energized.

M. Machines and Moving Parts

Employees working on normally moving parts of remotely controlled equipment shall be protected against accidental starting by proper tags installed on the starting devices, and by locking or blocking where practical. Employees shall, before starting any work, satisfy themselves that these protective devices have been installed. When working near automatically or remotely operated equipment such as circuit breakers which may operate suddenly, employees shall avoid being in a position where they might be injured from such operation.

N. Fuses

When fuses must be installed or removed with one or both terminals energized, special tools insulated for the voltage shall be used.

O. Cable Reels

Cable reels shall be securely blocked so they cannot roll accidentally.

431. Operating Routines

A. Duties of a Foreman

A foreman shall:

1. Adopt such precautions as are within his power to prevent accidents and to see that the safety rules and operating procedures are observed by the employees under his direction.

2. As far as possible, prevent unauthorized persons from approaching places where work is being done.

3. Prohibit the use of any tools or devices unsuited to the work at hand or which have not been tested or inspected as required by these rules.

B. Area Protection
1. Vehicular and Pedestrian Traffic
   a. Before engaging in work that may endanger the public, warning signs or traffic control devices, or both, shall be placed conspicuously to approaching traffic. Where further protection is needed, suitable barrier guards shall be erected. Where the nature of work and traffic requires it, a man shall be stationed to warn traffic while the hazard exists.
   b. In case openings or obstructions in the street, sidewalk, walkway, or on private property are being worked on or left unattended during the day, danger signals, such as warning signs and flags, shall be effectively displayed.

   Under these same conditions, at night, warning lights shall be prominently displayed and excavations shall be enclosed with protective barricades.
2. Employees
   If the work exposes energized or moving parts that are normally protected, danger signs shall be displayed and suitable guards erected to warn other personnel in the areas.
3. Crossed or Fallen Wires
   An employee finding any crossed or fallen wires which are or may create a hazard shall remain on guard or adopt other adequate means to prevent accidents, and shall have the proper authority notified.

C. Metal Tapes and Ropes
1. Metal measuring tapes, and tapes or ropes containing metal threads or strands, shall not be used above exposed energized parts, and shall not be brought closer than the following distances to exposed energized parts.

Voltage in volts	Distances in inches
8700 and less	30
8701–50,000	50

For voltages above 50 000, add 4 in for each 10 000 V of excess.

2. Care should be exercised when extending metal ropes, tapes, or wires parallel to and in the proximity of energized high voltage lines because of induced voltages.
3. When it is necessary to measure clearances from energized objects, only devices approved for the purpose shall be used.

432. Overhead Lines

All employees working on or with overhead lines shall observe the following rules in addition to applicable rules contained elsewhere in Section 43.

A. Checking Structures Before Climbing

1. Before climbing poles, ladders, scaffolds, or other elevated structures, employees shall determine, to the extent practical, that the structures are capable of sustaining the additional or unbalanced stresses to which they will be subjected.
2. Where there are indications that poles and structures may be unsafe for climbing, they shall not be climbed until made safe by guying, bracing, or other means.
3. Employees shall not climb poles where electric-supply wires or equipment are hanging below their proper level.

B. Position on Poles

When working on jointly used poles or structures, employees shall not climb or work above the level of the lowest electric-supply conductor exclusive of vertical runs and street light wiring.

EXCEPTION: This rule does not apply where communications facilities are attached above electric-supply conductors if a rigid fixed barrier has been installed between the supply and communications facilities.

C. Installing and Removing Wires or Cables

1. Precautions shall be taken to prevent wires or cables being installed or removed from contacting energized wires or equipment. Wires or cables which are not bonded to an effective ground and are being installed or removed in the vicinity of energized conductors shall be considered as being energized.
2. Sag of wire or cables being installed or removed shall be controlled to prevent danger to pedestrian and vehicular traffic.
3. Before installing or removing wires or cables, the strains to which poles and structures will be subjected shall be

considered and necessary action taken to prevent failure of supporting structures.

4. Employees should avoid contact with moving winch lines, especially near sheaves, blocks, and take-up drums.

5. Every employee working on or about equipment or lines exposed to voltages higher than those guarded against by the safety appliances provided shall assure himself that the equipment or lines on which he is working are free from dangerous leakage or induction or have been effectively grounded.

D. Setting, Moving, or Removing Poles in or Near Energized Electric-Supply Lines

1. When setting, moving, or removing poles in or near energized lines, precautions shall be taken to avoid direct contact of the pole with the energized conductors. Employees shall wear suitable insulating gloves or use other suitable means where voltage may exceed rating of gloves in handling poles where conductors energized at potentials above 750 V can be contacted. Employees performing such work shall not contact the pole with uninsulated parts of their body.

2. Contact with trucks, derricks, or other equipment which are not bonded to an effective ground and are being used to set, move, or remove poles in or near energized lines shall be avoided by employees standing on the ground, or in contact with grounded objects, unless the employee is wearing suitable protective equipment.

433. Underground Lines

Employees working on or with underground lines shall observe the following rules in addition to applicable rules contained elsewhere in Section 43.

A. Guarding Manhole and Street Openings

When covers of manholes, handholes, or vaults are removed, the opening shall be promptly protected with a barrier, temporary cover, or other suitable guard.

B. Ventilation and Testing for Gas in Manholes and Unventilated Vaults

1. The atmosphere shall be tested for flammable gas before entry.

2. Where flammable gases are detected, the work area shall be ventilated and made safe before entry.

3. Provision shall be made for an adequate continuous supply of air.

C. **Attendant on Surface**
   While personnel are in a joint-use manhole, an employee shall
   be available on the surface in the immediate vicinity to
   render assistance as may be required.

D. **Flames**
   1. Employees shall not smoke in manholes.
   2. Where open flames must be used in manholes, extra
      precautions shall be taken to ensure adequate ventilation.
   3. Before using open flames in an excavation in areas where
      combustible gases or liquids may be present, such as near
      gasoline service stations, the atmosphere of the excava-
      tion shall be tested and found safe or cleared of the
      combustible gases or liquids.

E. **Excavation**
   1. Cables and other buried utilities in the immediate
      vicinity shall be located to the extent practical prior to
      excavating.
   2. Hand tools used for excavating near energized supply
      cables shall be equipped with handles made of noncon-
      ductive material.
   3. Mechanized equipment should not be used to excavate in
      close proximity to cables and other buried utilities.
   4. If a gas or fuel line should be broken or damaged,
      employees shall:
      a. Leave the excavation open.
      b. Extinguish flames which could ignite the escaping gas
         or fuel.
      c. Notify the proper authority.
      d. Keep the public away until the condition is under
         control.

F. **Identification**
   1. When underground facilities are exposed (electric, gas,
      water, telephone, etc) they should be identified and shall
      be protected as necessary to avoid damage.
   2. Where multiple cables exist in an excavation, cables other
      than the one being worked on shall be protected as
      necessary.
   3. When multiple cables exist in an excavation, the cable to
      be worked on shall be identified by electrical means
      unless its identity is obvious by reason of distinctive
      appearance.
   4. Before cutting into a cable or opening a splice in a joint
      use manhole, the cable shall be identified and verified to
      be the proper cable.

G. **Maintaining Sheath Continuity**
When working on buried cable or on cable in manholes, sheath continuity shall be maintained by bonding across the opening or by equivalent means.

H. **Operation of Power Driven Equipment**
Employees should avoid being in manholes where power driven rodding equipment is in operation.

Index

INDEX

Italic type is used for C2 *rule, section, part, figure,* or *table* identification. The corresponding page numbers follow in upright type.

A

342

D

F

G

M

P

W

Procedure for Revising the National Electrical Safety Code

Preparation of Proposals for Amendment.

1.1 A proposal may be prepared by any:
- substantially interested person
- interested organization
- NESC (C2) Subcommittee
- member of the NESC (C2) Committee or its subcommittees

1.2 Five copies of the proposed shall be submitted to: Secretary, American National Standards Committee C2 (at the address listed in the time schedule for revision). Copies must be suitable for reproduction.

1.3 Each separate topic shall begin on a separate sheet or sheets, preferably printed or typed on one side only.

1.4 The proposal shall consist of
a. a statement, in NESC rule form, of the exact change, rewording or new material proposed
b. the name of the submitter (organization or individual as applicable).
c. supporting comment, giving the reasons why the NESC should be so revised
See SAMPLE submittal.

The C2 Secretariat will

(a) Acknowledge receipt of proposals for revision. (If the submitter does not receive an acknowledgement within 30 days of mailing his proposal, he should contact the secretariat).

(b) Distribute to each member of the appropriate C2 Subcommittee all of the proposals received, arranged in a coordinated sequence.

Subcommittee Recommendation

The C2 Subcommittee responsible will consider each proposal and take one or more of the following steps:

a) Endorse the proposal as received.

b) Prepare a proposed revision or addition for the NESC (this may be a coordination of several comments, or a committee consensus on a modification of a proposal.

c) Refer the proposal to a technical working group for detailed consideration.

d) Request coordination with other C2 Subcommittees.

e) Recommend rejection of the proposal, for stated reasons.

For each item, prepare a subcommittee voting statement, accompanied by all member's statements concerning their votes (cogent reasons required for negative votes). Steps c) and d) are intended to result in an eventual proposal of category (b).

Action under steps c) or d) shall be completed and reported to the subcommittee before the end of the 5 month public review period, if the item is to be included in the upcoming revision.

Preprint of Proposals

The C2 Secretariat will organize and publish a preprint of proposed C2 revisions including

- The original proposal as received from the submitter
- The recommendation of the subcommittee with respect to the proposal (including a voting statement and subcommittee member's statements)

This preprint will be distributed to all members of C2 Subcommittees and the representatives of the organizations comprising the C2 Committee. Copies will be available for sale to other interested parties. The preprint will carry information on how to submit comments on the proposals and the final date for such submissions.

Final Processing of Proposed Revisions and Comments (following the 5 month review period).

5.1 The C2 Secretariat will organize and distribute for Subcommittee consideration all comments received during the 5 month review period.

5.2 The preprint and the comments received will be reconsidered by the subcommittees
a) If there is need for technical consideration or resolution of differing or conflicting points of view, the subcommittee shall refer the problem to a working group of the subcommittee for a proposed resolution.
Each working group shall provide to its parent subcommittee, recommendations on matters considered as a result of subcommittee referrals under items 3c and 5.2a.
b) If there is a consensus, the subcommittee may recommend adoption or rejection of the proposal
Each subcommittee shall prepare a report showing its proposed revisions and all items to be held on the docket, together with a plan for their disposition.

6. Final Approval

6.1 Based upon the Subcommittee reports the C2 Secretariat shall prepare a draft revi of the NESC (C2) and distribute this to the

a) C2 Committee for approval by a 6 week letter ballot

b) American National Standards Institute, Board of Standards Review for concur 60 day public review.

6.2 C2 Committee negative voting shall be by identified items. Those items which re consensus and have no technical comments during the public review period shall come the revisions for the next edition of C2.

6.3 Those items on which a consensus cannot be reached, or on which substantial techn comments are received during the 60 day public review period shall be referred to appropriate subcommittee for resolution.

6.4 ANSI Approval

As soon as practicable after the close of the ANSI Public Review Period, the C2 proved revisions shall be submitted to ANSI for approval of the proposed new edi

---

Request for Change
National Electrical Safety Code

Rule 214B2 Revise rule to read:

Lines temporarily out of service shall be maintained for operation under the rules.

Submitter: J. D. Doe, X Y Z Electric Co

Supporting Comment: A utility cannot comply with the present rule. Once an accident has curred it is usually prima facie evidence that a hazard was created. If the existing rules are s factory for lines in service, they should be satisfactory for lines temporarily out of service.

Time Schedule for the Next Revision
National Electrical Safety Code

1981 Sept 15   Final date for receipt of proposals from the public for revision of the 1981 ed tion National Electrical Safety Code, preparatory to the publication of a 198 edition. Proposals should be forwarded in the prescribed form to:

Secretary
American National Standards Committee C2 on NESC
IEEE Standards Office
345 East 47th Street
New York, NY 10017

Nov 9 to Dec 11   NESC Subcommittees consider proposals for Code changes and prepare th recommendations.

1982 Feb 1   Preprint of Proposed Amendments for incorporation into the 1984 Edition of th National Electrical Safety Code published for distribution to the C2 Commit and other interested parties.

Feb to June 30   Period for study of proposed amendments and submittal by interested parties recommendations concerning the proposed amendments. Submit recommen tions to the Secretary, ANSC C2 on NESC, at the address above.

Sept 7-30   Period for NESC Subcommittee Working Groups and NESC Subcommittees reconsider all recommendations concerning the proposed amendments and p pare final report.

1983 Feb 1   Proposed revision of NESC, ANSI C2, submitted to C2 Committee for let ballot and to the American National Standards Institute for concurrent pub review.

May 1   C2 Committee approved revisions of NESC submitted to American National Sta dards Institute for approval as an ANSI standard.

July   Publication of the 1984 Edition of the National Electrical Safety Code

366